THE YOUNG OXFORD BOOK OF

Astronomy

THE YOUNG OXFORD BOOK OF

Astronomy

Simon and Jacqueline Mitton

Oxford University Press

CONTENTS

Oxford University Press

Oxford New York
Athens Auckland Bangkok Bombay
Calcutta Cape Town Dar es Salaam Delhi
Florence Hong Kong Istanbul Karachi
Kuala Lumpur Madras Madrid Melbourne
Mexico City Nairobi Paris Singapore
Taipei Tokyo Toronto
and associated companies in
Berlin Ibadan

Copyright © 1995 by Simon and Jacqueline Mitton
The authors can be contacted by electronic mail at
smitton@cup.cam.ac.uk.

Published by Oxford University Press, Inc.

200 Madison Avenue
New York, New York 10016

Library of Congress Cataloging-in-Publication Data
Mitton, Simon, 1946-
Astronomy / Simon and Jacqueline Mitton.
p. cm.—(Young Oxford books)
Includes index.
1. Astronomy—Juvenile literature
[1. Astronomy.]
I. Mitton, Jacqueline. II. Title. III. Series
QB46.M63   1995
520—dc20   95-7015
                            CIP
                            AC

ISBN 0-19-521169-3 (trade ed.)
ISBN 0-19-521168-5 (lib. ed.)

9 8 7 6 5 4 3 2 1
Printed in Mexico.

Front cover The Sun and planets of our solar
system.

1
THE CHALLENGE OF ASTRONOMY

2
THE SOLAR SYSTEM

3
The Sun and the Stars

4
Galaxies and the Universe

INTRODUCTION

Look up on a clear, dark night and you will see the sky is full of twinkling specks of light. What are these stars and planets? How far away are they? Can we learn more about them? People have asked themselves questions like these for thousands of years. And since you have opened our book, no doubt you are looking for the answers, too.

Before telescopes were invented around 400 years ago, it was very difficult for astronomers to learn much. Today, telescopes and spacecraft are collecting amazing pictures and information about the planets, stars, and galaxies. In this book we tell you about the different ways in which astronomers are exploring the universe now. Then we set out on a tour of space, starting with the planets. Next we stop off at our own star, the Sun, before looking at the huge variety of stars in our own Galaxy, the Milky Way. Beyond the Milky Way is a whole universe of galaxies to explore. And if it is clear tonight, you can make a start on your own journey of discovery in space by finding your way around the constellations with our easy-to-follow star charts. Have fun!

Simon Mitton
Jacqueline Mitton

THE CHALLENGE OF ASTRONOMY

We live on planet Earth, orbiting the Sun in company with the Moon. The Sun is one star among billions in the Milky Way Galaxy. There are other galaxies so far away that their light takes billions of years to reach us. Using telescopes and spacecraft, astronomers make sense of our view of the Sun and the starry night sky.

EXPLORING THE UNIVERSE

Astronomy is about exploring and understanding the universe. Outer space contains matter in many shapes and forms: dust and gas, planets and comets, our Moon and other moons, the Sun and other stars, the Milky Way and other galaxies.

This book tells you what astronomy is and what astronomers can find out about the universe. We shall journey to the planets of our solar system, look at how stars work, and find out about the beginning and end of the universe.

Professional astronomy today is a major science that investigates the properties of the planets, stars, and galaxies in the universe. Astronomers want to know the history of the universe and how the different kinds of objects in it came to be formed. They measure distances from Earth to nearby planets and to the farthest galaxies. Astronomers try to discover the inner workings of the Sun and stars by using mathematics and physics. They use very expensive and sophisticated equipment, such as large telescopes, spacecraft, and the fastest computers in the world.

But astronomy is not just for professional scientists. Without buying any equipment at all, you can get started on astronomy, perhaps even tonight. With inexpensive binoculars, you can explore the craters of the Moon, clusters of stars, and the Milky Way.

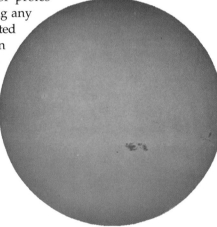

▽ The Sun is a star, and its diameter is 109 times greater than Earth's diameter.

◁ Earth photographed from the *Apollo 17* spacecraft, 25,000 miles from Earth. The entire continent of Africa is visible.

◁ This spiral galaxy (NGC 6946) is similar to the Milky Way Galaxy, which includes the Sun.

▽ Comet West, showing a tail of gas and dust streaming away from the comet's head. The images of stars were blurred as the camera followed the motion of the comet.

WHAT IS THE UNIVERSE?

In science the word *universe* has a special meaning. The universe is the largest possible volume of space, together with all the matter and radiation within it, that can affect us in any way. There is only one observable universe from our point of view. There may be other universes, and the universe we observe may extend without limit. The volume of space we observe, which now stretches out to about 17 billion light-years in every direction, is "our" universe.

△ The surface of the planet Venus is covered with many volcanic craters and flows of lava.

◁ NGC 3932 is a cluster of young stars.

What can we see tonight?

Let us start with the greatest show on Earth, which you can see free of charge: the starry night sky. Try to look at the night sky for half an hour or so next time there are no clouds. Wear warm clothes and have a flashlight handy.

On a dark night, away from city lights, you can see thousands of stars and the misty path of the Milky Way. As night progresses, the patterns of stars slowly wheel from east to west. Try to pick out some star patterns that are easy to remember. See where the brightest stars are and see what shapes and patterns they make.

If it is not too late, look again an hour or so later. The star patterns have moved somewhat and are farther to the west. This is because the Earth is rotating. You have already made an important scientific observation by observing the evidence that the Earth is spinning in space.

▷ The constellation Orion (the Hunter) is visible from almost anywhere on Earth in the evenings from about November to March.

From myth to modern science

For thousands of years people have looked at the night sky using just their eyes. The earliest astronomers gave names to many of the star patterns as well as to individual stars. Different societies each identified their own particular star patterns and had their own names for them. Astronomers still use many of the names chosen by Greek observers of the night sky over 2,000 years ago.

It is important to understand the difference between astronomy and astrology. Astronomy is a science based on careful observation. Astrology is a form of popular entertainment based on superstition. Astrologers' claims to predict future events are not now believed by the majority of scientists but, centuries ago, astrology was taken more seriously and was the reason many astronomical observations were made.

Planets, stars, and galaxies

We live on planet Earth in orbit around a star we call the Sun. There are nine major planets circling the Sun. Seven of them have moons. The Sun, the planets, and their moons make up the solar system, along with huge numbers of asteroids and comets, and quantities of dust. Spacecraft have visited or flown near all of the planets in our solar system except Pluto, which is the most distant.

The most important difference between stars and planets is this: stars shine by their own light, but planets shine by reflecting light from the Sun. Stars make their light from nuclear energy deep inside. They are extremely hot balls of glowing gas. Normal stars, such as the Sun, are much larger than planets: the Sun's diameter is over 100 times greater than Earth's. Planets, moons, and asteroids shine by reflecting sunlight. Planets are either rocky, like Earth, or giant balls of cool gas, like Jupiter.

New stars form inside clouds of gas and dust in space. The Sun was born in a gas cloud about 5 billion years ago and is part of an enormous family of billions and billions of stars that together make up the Milky Way Galaxy. The stars and gas clouds form a beautiful spiral pattern. Our Galaxy is enormous. If you could travel at the speed of light, it would take about 100,000 years to cross.

When you look at the night sky, almost every point of light you see is a star in our Galaxy. Sometimes you can see one or more of the planets. As they travel around the Sun, they move through the star patterns from our viewpoint on Earth. Five of them are visible without a telescope if you know when and where to look. The brightest and easiest to spot are Venus and Jupiter. Venus can often be seen in the eastern sky just before dawn or in the

10

Jupiter   Saturn   Uranus   Neptune   Pluto

Sizes

◁ The lower part of the diagram shows the sizes of the major planets compared with the Sun. The upper part shows their orbits around the Sun. Mercury, Venus, Earth, and Mars are small, rocky planets that together make up the inner solar system. Jupiter, Saturn, Uranus, and Neptune are huge planets made of gas. All of the planets, except Pluto, have been visited or passed by spacecraft carrying cameras and scientific equipment.

Orbits
1. Sun
2. Mercury
3. Venus
4. Earth
5. Mars
6. Jupiter
7. Saturn
8. Uranus
9. Neptune
10. Pluto

western sky after sunset. What looks like an especially bright "star" shining with a steady yellow light may be the planet Jupiter.

Other galaxies vary widely in shape and size, but they have one thing in common: all are families of stars, gas, and dust. The most distant objects that astronomers can see are very bright galaxies.

The universe

All the galaxies, with their stars, planets, and dust, together make up the universe. There are many galaxies but only one universe. The universe includes everything we can see through telescopes and everything that can affect us in any way.

The universe is so large that it is impossible to imagine what it is like as a whole. It has taken light rays about 17 billion years to reach us from the most distant parts of the universe. Inside our enormous universe there are thousands of billions of galaxies, each containing billions of stars.

Astronomers think that our universe began in a colossal explosion, about 17 billion years ago. This event is known as

the Big Bang, and it took place at the beginning of time. Everything that is now in the universe formed originally from the hydrogen and helium that were made in the Big Bang.

You might wonder if there is anything outside the universe. Are there more universes? Did a different universe exist before the Big Bang? We simply do not know the answers to all these questions. Science only tells us about the universe we live in.

Seeing farther, seeing better

Telescopes are the most important tools for astronomers. With a telescope an astronomer can detect much fainter objects than those you can see with the naked eye. Astronomical telescopes can be designed for studying the X-rays, radio waves, infrared and ultraviolet light from space. All of these radiations are invisible to our eyes, but astronomers need to observe them in order to get a full picture of what the universe looks like.

Before the 17th century, astronomers had to manage without telescopes. The main aim of ancient and medieval

▷ Galileo Galilei made the first astronomical observations with a telescope.

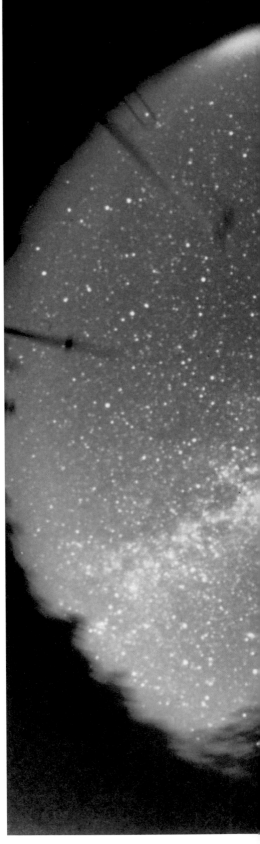

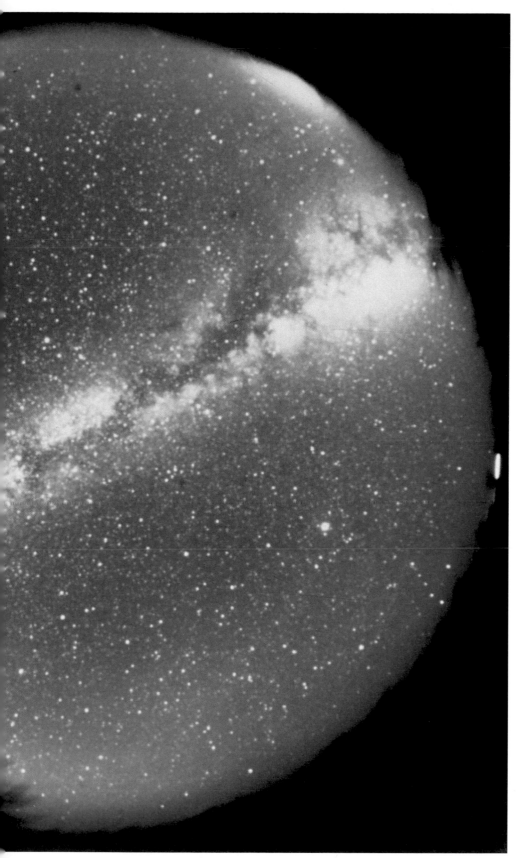

△ This full-sky view of the Milky Way was taken from Mount Lemmon, 10,500 feet high in Arizona. Note the dark streak of dust running through the central plane of the Milky Way.

astronomy was to follow the planets across the sky so that their future positions could be predicted. The only instruments were simple ones for measuring positions in the sky.

Galileo Galilei (1564–1642) introduced telescopes to astronomy. Although he did not invent the telescope, he was probably the first person to use one for astronomy and record his observations. He started using his telescope in July 1609. Among his most important discoveries were four of the moons of Jupiter, which looked like a miniature solar system, and that the faint light of the Milky Way comes from countless millions of stars. Galileo changed the science of astronomy dramatically by showing that the Sun is at the center of our solar system and by hinting at the vast size of the starry universe.

LIGHT-YEARS

You can measure the distance between two cities in miles, but you might also say it takes "two hours to drive." From New York to London the distance is 3,428 miles, but most air travelers describe it as a "seven-hour flight." Out in space the distances are enormous. Astronomers often describe them in terms of the travel time for light.

The speed of sound in air is about 760 miles per *hour*, and some aircraft can fly faster than this. The speed of light is about 186,281 miles per *second*. Nothing can travel faster than light. A spaceship flying from the earth to the sun would take 17 years if it traveled at the speed of sound. Light from the Sun takes just 8 minutes to cross the same distance. This Sun–Earth distance of 93 million miles is "eight minutes flying time" for a ray of light.

In one year light travels nearly 6 million million miles. The distance light travels in one year is known as a light-year. You can work out the distance: calculate how many seconds there are in a year, then multiply the result by 186,281. The light-year is used instead of trillions of miles for describing distances in the universe.

OBSERVING THE UNIVERSE

Astronomers use telescopes and space probes to study everything beyond the Earth. Light and other radiations from stars and galaxies carry information about their speeds and many other properties. With practice you can see many different astronomical objects with the naked eye or with binoculars.

Without a telescope, you can see only a few thousand stars. Planets and the stars are very far away, so their light is very dim by the time it reaches us. To make up for the faintness of the light, we need to "collect" as much of it as possible. With astronomical telescopes, the light from billions of stars comes into view.

An astronomical telescope works by collecting the faint light from a distant planet, star, or galaxy and concentrating it on the eye of an observer. The light collector is either a lens (called the objective lens) at the front of a tube or a curved mirror at the bottom of a tube. A telescope with an objective lens is called a refractor; one with a mirror is called a reflector.

The light-gathering power of an astronomical telescope and its ability to distinguish fine detail depend on the size of its main lens or mirror. The main lens or mirror of a telescope has a much greater area than the pupil of your eye. The light falling on this larger "collecting area" means that a much greater amount of light is funneled to the observer. An inexpensive pair of binoculars lets you see tens of thousands of stars, and with a small telescope millions come into view.

The method used by the telescope to detect the light is important as well. The more sensitive the detector is to light falling on it, the fainter the objects the telescope can see. The detector can be a human eye, but only amateur astronomers observe this way these days.

To look at the image in a telescope directly with the naked eye, you need to use an eyepiece. Eyepieces are usually made of two or more small lenses fixed in their own little tube. You can interchange the eyepieces on a telescope to alter the magnification or the area of sky you can see at a time. Very high magnification is not important for most optical observations. In the case of the planets, for example, too much magnification produces a blurred and shaky image. When using a small telescope, do not use a strongly magnifying eyepiece because the image will then look rather blurred.

Professional astronomers use photographic films and plates, or electronic devices. With electronic recording equipment, a computer is needed to create an image on a screen and to store the data as a collection of numbers. Magnification does not matter to professional astronomers because once a photograph or computer image has been produced it can be enlarged later if necessary.

Refracting telescopes

Galileo used lenses in his telescopes to capture light and focus it into an image he could look at with the naked eye. In this kind of telescope, called a refractor, the lens at the front—the objective lens—forms the main image near the bottom end of the telescope tube. Ordinary binoculars are made of two refracting telescopes mounted side by side.

The largest refractor in the world, at the Yerkes Observatory in Wisconsin, has a lens 40 inches in diameter. Refracting telescopes are no longer used very much in professional astronomy. Astronomers want to collect as much light as possible and it is much easier to build really large telescopes with mirrors. A lens more than about three feet across would be too thick and too heavy to be of any use in a telescope. Many amateurs like to use smaller refractors, particularly for observing.

▷ Binoculars are two miniature refracting telescopes mounted side by side. Small prisms inside each tube reflect the light in such a way as to keep the length short and to turn the image right side up. Binoculars are ideal for getting started with astronomical observation.

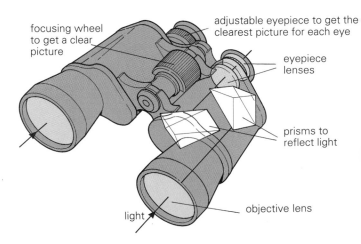

focusing wheel to get a clear picture

adjustable eyepiece to get the clearest picture for each eye

eyepiece lenses

prisms to reflect light

light

objective lens

◁ The observatory at La Palma, Canary Islands, showing the dome of the William Herschel Telescope (left) and a solar tele-scope in a tower (right). This is the largest astronomical observatory in Europe.

◁ The 4.2-meter William Herschel Telescope at La Palma, Canary Islands. The telescope is equipped with many different scientific instruments designed to study the light from stars and galaxies. The reflecting mirror is 14 feet in diameter.

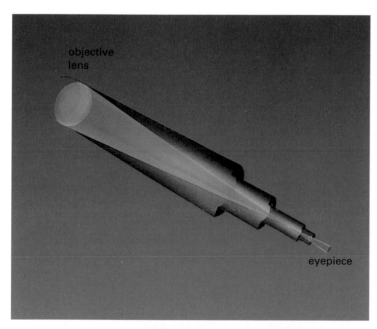

objective lens

eyepiece

◁ A refracting telescope has an objective lens, to form the image, and an eyepiece for magnifying the image.

Reflecting telescopes

The idea of using a mirror in a telescope instead of an objective lens was first put forward by James Gregory (1638–75) in 1663. Light can be collected and focused by a concave mirror instead of a lens. The main image is formed in front of the mirror (at a place called the prime focus), so a reflecting telescope needs at least one extra mirror to send the image to a more convenient place.

In small telescopes the light is often deflected through the side of the telescope tube by a small flat mirror held inside the tube. Reflectors like these are called Newtonians after Sir Isaac Newton (1642–1727), who made the first success-ful reflecting telescope in 1668. Another arrangement, often used on research tele-scopes and on some kinds of amateur telescopes, is to have a small curved mirror near the top of the tube, aiming the light back down again and through a hole in the center of the main mirror. Various other ingenious arrangements are possible on very big telescopes. In some cases, if the telescope is large enough, an observer can sit in a cage suspended at the top of the telescope tube! Reflectors, espe-cially large ones, do not need closed tubes. Usually, there is just a framework of metal bars to hold the mirrors and instruments in place. The world's largest reflector, the Keck Telescope, is in Hawaii. Its main mirror consists of 36 sections, which are accurately controlled by computers.

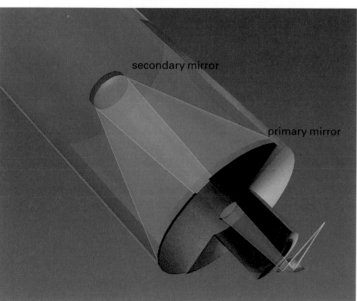

secondary mirror

primary mirror

◁ This reflecting telescope has a concave primary mirror to form the image. A secondary mirror, also curved, reflects the light to the eyepiece. The eyepiece can be replaced by a camera to carry out photography.

Photography and imaging

Observing directly with the naked eye is fun, but professional astronomers and many amateurs, too, want a permanent and accurate record of what they see, so they use photographic plates or special electronic devices. These light detectors also have another advantage. The images of faint objects can be built up in a time exposure over many minutes, or even hours, provided the telescope can track the moving stars very accurately. When you look through a telescope for the first time, you will probably be very disappointed. Almost all the photographs you see in books (including this one) are time exposures made on large telescopes. Most of the detail in the pictures cannot be seen by looking through the telescope.

Today most researchers use electronic light detectors. These can register individual blips of light energy (known as photons). Photographs are still used for some purposes, but electronic detectors are much more sensitive than photographs.

They use materials sensitive to light. When photons fall on them, they produce a tiny electric current. This type of detector can be used to measure the brightness of a star.

Charge-coupled devices (or CCDs) are like a checkerboard of light detectors. They build up an image by recording which square on the "board" each photon hits. Each of the squares is called a picture element, or "pixel," for short. On astronomical images that have been magnified greatly, you can see the pattern of pixels. Detectors of this kind are used in hand-held video cameras. A CCD camera is about 50 times more sensitive than an ordinary photographic camera.

Color your own images

What are the real colors of the stars and planets? We think of "real" colors as being the ones we see with our eyes, but eyes are just one kind of light detector and not everyone sees colors in the same way. Photographs and electronic light detectors can record colors more accurately.

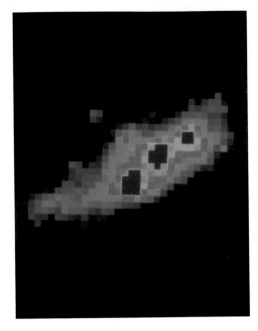

△ This picture of a galaxy was taken by a CCD camera. It shows the individual squares, known as pixels, of the image.

Wavelength scale

| atom 1 ten-millionth of a mm | bacteria 1 ten-thousandth of a mm | dust particles 1 hundredth of a mm | pinhead 1 mm | child 1 m |

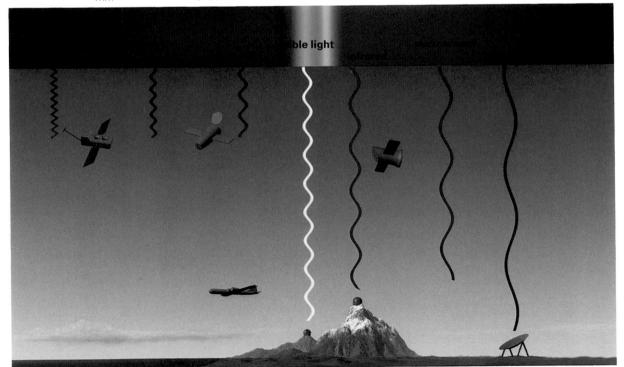

ble light

◁ The electromagnetic spectrum, from radio waves to gamma rays. Visible light is just one example of this form of radiation. Radio waves have the lowest energy, and gamma rays, the highest. Infrared radiation is also known as heat radiation. Radio waves and light can be studied using telescopes at ground level. For other forms of radiation, telescopes must be put on high mountains or sent into space on orbiting satellites.

When an image is stored in a computer as a set of pixels, it can be shown on the screen or printed in any chosen colors. If they are not the natural ones, they are said to be "false colors." You can see the effects just by altering the color balance on a TV set. False colors help bring out the contrast in an image. Pictures showing how an object would look in radiation invisible to our eyes, such as infrared light, X-rays, or radio waves, always have to be coded in false color.

Beyond the rainbow

Light is a form of energy called electromagnetic radiation. It can travel through empty space and is in the form of individual wave packets called photons. The waves in packets of visible light are tiny ripples less than a millionth of a millimeter long. When light is split into a spectrum of colors, as in a rainbow or after it has gone through a glass prism, you can see a spread of the different wavelengths. Violet light has about half the wavelength of red light.

But light is not the only form of electromagnetic radiation. The whole electromagnetic spectrum extends beyond rainbow colors, past the violet to much shorter wavelengths and past red to much longer wavelengths. At the longer wavelengths are infrared radiation, microwaves and radio waves. Going to shorter wavelengths past violet are ultraviolet radiation, X rays, and gamma rays.

Astronomers study the whole spectrum from end to end. Not every kind of radiation gets through the air to the ground, however. Gamma rays, X rays, and ultraviolet radiation have to be viewed from satellites orbiting the earth above the atmosphere.

By observing how much radiation we receive at particular wavelengths from an object such as a star or glowing gas cloud, astronomers can work out details of that object's density and temperature, its motion, and its chemical composition. Splitting up electromagnetic radiation by wavelengths in order to study it is called spectroscopy—looking at the spectrum.

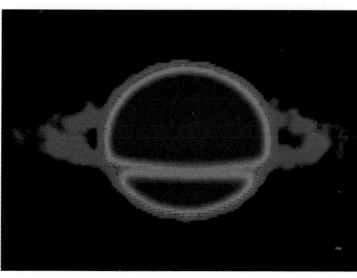

◁ Two views of Saturn in visible light (above) and in false color (below). In the first, sunlight reflecting from the planet's cloud tops and rings lights the image. In the second, color is used to represent the strength of radio emission with a wavelength of two centimeters (about one inch).

▷ The rainbow colors, or spectra, of stars can tell astronomers what stars are made of and how hot they are.

GETTING STARTED AS AN ASTRONOMER

Suppose you want to see more of the stars and planets. What telescope should you get and what do you do to get help? Any experienced amateur will tell you that the kinds of inexpensive telescopes you find in ordinary shops are usually a great disappointment. Remember that magnification is not important! You need good quality lenses and mirrors and a firm tripod. If you want a telescope, it is best to go to a shop that specializes in astronomical telescopes.

The next best thing is to get an inexpensive pair of binoculars. Many amateur astronomers choose to sweep the sky with binoculars, searching for interesting objects and new comets. Even if you have neither a telescope nor binoculars, you can still enjoy finding your way around the sky by using star maps.

In many areas, there are local astronomical societies, and members are almost always willing to help newcomers.

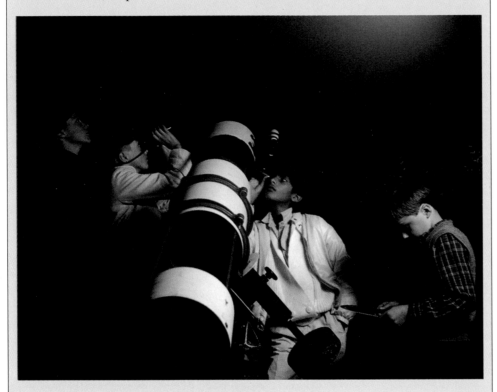

Radio astronomy

Radio astronomy started in the 1940s after the first cosmic radio signals were detected by chance. Many different objects, from the Sun to galaxies and even the universe itself, emit radio waves. Fortunately, radio waves can penetrate through the atmosphere, so large radio telescopes can be built on the ground. They can operate day and night whatever the weather, since radio signals even go through clouds.

A radio telescope works in much the same way as an optical telescope: it collects radiation, brings it to a focus on a detector tuned in to a chosen wavelength, then converts it to a form in which a contour map or false-color image of the sky can be made showing the strength of radio signals at that wavelength. Because radio waves are used so much for communications, some wavelength bands are specially reserved and kept clear for radio astronomers.

The most familiar kind of radio telescope uses a large dish as the antenna or aerial to collect the radio waves. It acts in exactly the same way as the curved mirror of a reflecting optical telescope, and the larger it is the fainter the signal it will be able to pick up.

Because radio waves are much longer than light waves, the size of a radio telescope needs to be scaled up in proportion if it is going to capture as much detail in its maps as an optical telescope can achieve in its images. This would require dishes many miles wide, which is obviously

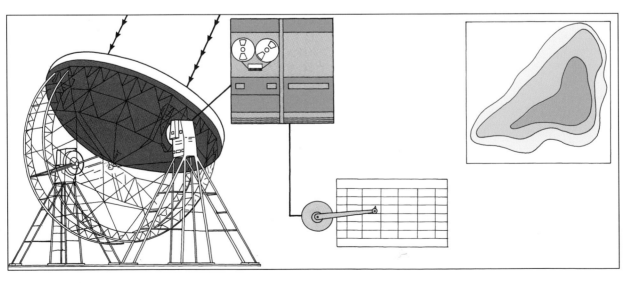

◁ The dish of a radio telescope focuses the radio waves onto a detector, where the signal is passed to amplifying and recording equipment and then to a device that records the signal in graph form. This can be used to build up a contour map of a region of the sky.

▷ The Lovell radio telescope at the Nuffield Radio Astronomy Observatory near Manchester, England.

impossible. Radio astronomers get around the problem by linking a number of smaller dishes and feeding their simultaneous observations into a computer. Some radio observatories have a whole set of dishes on one large site. There are also linkups between dishes that are far apart, in different countries or even different continents!

Space telescopes

The Earth's atmosphere is vital to our existence. It contains the oxygen we breathe, acts like a blanket to keep the surface warm enough to sustain life, and filters out harmful ultraviolet radiation from the Sun. But for astronomers, the atmosphere is a mixed blessing. Space telescopes orbiting high above the atmosphere are needed to observe outside the visible and radio bands of the spectrum, and a limited range of the infrared. The visible star images we can see are changed by the air from sharp, steady points of light to shimmering circles. The twinkling of the stars, caused by unsteadiness in the air, makes it difficult to get sharp astronomical pictures from the ground, though some ingenious methods of combating the problem have been invented. Then there is cloudy weather to contend with. Another growing problem is light pollution. City and road lights scattering in the air cause the whole atmosphere to glow faintly. This background light greatly interferes with astronomical observations.

Part of the answer is to build major observatories on remote mountaintop sites. There is even one observatory built into a high-flying aircraft, but for some kinds of observations, only space telescopes will do. One of the best known is the Hubble Space Telescope, run jointly by the United States and a partnership of European countries. It has a main mirror 7.8 feet wide and was designed to make many of its observations in ultraviolet as well as in visible light. When launched in 1990, it was intended to operate for around 15 years. At the end of 1993, a team of astronauts repaired the Hubble Space Telescope by replacing several parts that were damaged or faulty.

Many other space telescopes have been operated successfully since the 1960s, and more are always being planned. Each is specially designed to probe different parts of the spectrum and typically works for a few months or a few years. These automated telescopes are programmed to operate on radioed commands from astronomers at control stations on the ground.

Planetary probes

The stars and galaxies are so far away that we have no hope at the present time of sending spacecraft to take a close look, but in the solar system it is perfectly feasible to explore by robot. Not all missions are successful, but every planet (except Pluto), the Moon, Halley's Comet, and some asteroids have all had spacecraft fly close by, go into orbit around them, or land on their surface. As a result, our knowledge of the planetary system has increased enormously since the 1960s.

Close-up images of the planets and their moons are relayed to earth by radio signals from TV cameras on board. Generally, there are other instruments too, taking measurements of the space environment.

Orbiters have been put in place around Mars and the Moon, making it possible to map their entire surfaces over a period of months. The *Magellan* orbiter around Venus has also mapped the whole planet, but by a different method: radar. Venus is entirely covered by opaque clouds, so it is impossible to see the surface. Radar bounces radio signals from the surface to build up a picture of the surface features.

The Moon is the only place beyond the Earth to be visited by humans. There were six successful landings in the Apollo program of the United States between 1969 and 1972. The astronauts were able to bring back samples of moon rock for geologists to study. More lunar samples were collected by a Soviet probe in 1976. Unmanned spacecraft have landed successfully on Venus and Mars.

Two of the most productive missions for planetary astronomers were *Voyagers 1* and *2*, launched by the United States in 1977. Both traveled to Jupiter and Saturn and returned stunningly beautiful images of these giant planets, as well as many of their moons. *Voyager 2* also flew by Uranus in 1986 and Neptune in 1989, successfully returning still more remarkable images.

◁ The Hubble Space Telescope is the most productive telescope ever placed in space. It has transformed planetary science and lets astronomers see very fine detail in distant galaxies.

THE BUILDING BLOCKS OF THE UNIVERSE

Everything in the universe—galaxies, stars, planets, living things on Earth—is made from a number of basic chemical elements. The tiniest particle of a chemical element is called an atom. Atoms are made of subatomic particles called protons, neutrons, and electrons. Groups of atoms combine into molecules to form the great variety of materials in the world.

Astronomers want to know how galaxies form, why stars shine, and what the planets are made of. To answer these and many other questions about the universe, we need to understand what matter is like. Planet Earth and even our own bodies contain exactly the same basic elements as the most distant galaxies. Therefore, what scientists have learned in laboratories about the tiny particles that make up matter is very important in astronomy.

Suppose we have an extremely powerful microscope. What happens as we close in on, say, a lump of iron? As the magnification is increased, there is not much change to start with. Small bits of the iron look more or less the same as the whole piece. But with magnifications of millions and millions and millions of times, the iron appears to be made of misty balls touching each other. The most powerful microscopes in the world are just capable of detecting the smallest possible pieces of iron.

In about 400 B.C., the Greek scholar Democritus suggested that all matter is composed of minute indivisible particles, which he called atoms, from the Greek word meaning "uncuttable." He could not prove his idea, but we know now that he was mostly right—except that atoms can be split into even smaller particles.

Inside an atom

A hundred million atoms in a row would stretch for only one inch. But each atom has inside it a nucleus thousands of times smaller still. Imagine a model of an atom with a nucleus the size of a pinhead. The whole atom would be about 30 feet across.

An atomic nucleus is made up of two kinds of particles—protons and neutrons. They are held together by strong nuclear forces. Protons carry a tiny amount of positive electric charge, but neutrons have no charge. Both have about the same mass.

Around the nucleus, a cloud of electrons forms the outer part of the atom. Each electron carries the same quantity of electric charge as a proton, but it is of the opposite (negative) kind. The force of attraction between the positive charge on the nucleus and the negative charge on the electrons holds them in place in the atom. Electrons are much lighter than protons and neutrons. It takes nearly 2,000 electrons to make up the mass of a proton. This means that almost all the mass in an atom is concentrated in its nucleus—and most of an atom is made of empty space!

A normal atom contains the same number of electrons as protons, so the total electric charge of the atom is zero. However, if an atom gets hot, or collides with another particle, one or more of its electrons can get torn away, leaving the atom with a positive charge. This happens frequently to atoms in stars and interstellar space.

Atoms and elements

The simplest of all atoms has just one proton for its nucleus and one electron. Atoms like this make up the gas hydrogen, which is the lightest of the chemical elements. Throughout the universe, there is far more hydrogen than anything else.

Different chemical elements are built by having different numbers of protons in the atomic nucleus. For example, carbon has 6 protons; iron, 26 protons; and lead, 82 protons. Neutrons are needed in a nucleus as well to make it stable (apart from the

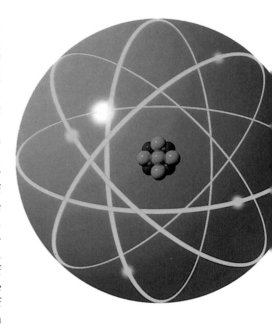

△ An atom of the element beryllium has in its nucleus four protons carrying positive electric charges and five uncharged neutrons. Four negatively charged electrons in a cloud around the nucleus make the total charge on the atom zero. The size of the nucleus is greatly exaggerated in the diagram. It would be an invisible pinprick compared with the size of the electron cloud.

special case of hydrogen). Carbon has 6 neutrons, iron 30, and lead 126. Lead is denser than iron and feels so heavy because it has such a large number of neutrons and protons in each of its atoms. Each element has been designated with a one- or two-letter abbreviation, such as H for hydrogen, He for helium, and C for carbon.

Some very heavy nuclei are not stable whatever the number of neutrons they contain. They fire out particles to convert themselves into lighter, stable elements. Elements that do this are called radioactive. Uranium and radon are two radioactive elements found naturally on the Earth. All the elements with 84 or more protons are radioactive. In total, 90 different chemical elements occur naturally on our planet.

Molecules

Atoms of the chemical elements are the main building blocks of our universe, but most of the world around us is not made of simple elements. There is a huge variety of different materials because atoms of different elements can link together to form molecules.

The simplest molecules are two or three atoms sharing some of their electrons. For example, water molecules contain two atoms of hydrogen and one of oxygen. The gas carbon dioxide contains one atom of carbon and two of oxygen. In scientists' shorthand, they are represented as H_2O and CO_2. At the other end of the scale, molecules can be made of hundreds or thousands of atoms. Very large, complicated molecules are the basis of all living things.

Molecules are easily torn apart by high temperatures, so they are found only in the coolest of stars. But in the icy depths of space, astronomers have discovered giant clouds containing numerous different molecules.

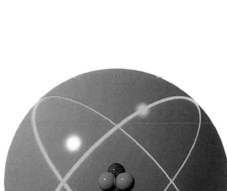

◁ An atom of the simplest chemical element, hydrogen, has a single proton for its nucleus and one electron. It is the only element with no neutrons in its nucleus.

▷ Atoms of the second lightest element, helium, have two protons and two neutrons in their nuclei and two orbiting electrons.

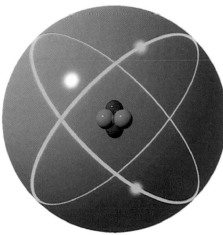

◁ In the nucleus of an oxygen atom, there are eight protons and eight neutrons. A cloud of eight electrons surrounds the nucleus.

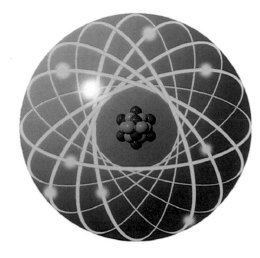

SOME FAMILIAR CHEMICAL ELEMENTS			
Name	*Symbol*	*Protons*	*Neutrons*
Hydrogen	H	1	0
Helium	He	2	2
Carbon	C	6	6
Nitrogen	N	7	7
Oxygen	O	8	8
Neon	Ne	10	10
Sodium	Na	11	12
Magnesium	Mg	12	12
Aluminium	Al	13	14
Silicon	Si	14	14
Phosphorus	P	15	16
Sulphur	S	16	16
Chlorine	Cl	17	18
Potassium	K	19	20
Calcium	Ca	20	20
Iron	Fe	26	30
Nickel	Ni	28	30
Copper	Cu	29	34
Arsenic	As	33	42
Krypton	Kr	36	48
Silver	Ag	47	60
Tin	Sn	50	70
Tungsten	W	74	110
Platinum	Pt	78	117
Gold	Au	79	118
Mercury	Hg	80	122
Lead	Pb	82	126
Radon	Rn	86	136
Uranium	U	92	146

OUR PLACE IN THE UNIVERSE

Earth is the third planet from the Sun, orbiting at an average distance of 93 million miles. The Sun is about 25,000 light-years from the center of our home galaxy, the Milky Way. The nearest large galaxy to our own, the Andromeda Galaxy, is 2 million light-years away.

When you take a trip in a car, train, or aircraft, you know when you are moving. You have the sensation of things whistling by, and the going is not always smooth. But when you are standing still on the ground, how do you know whether you are really moving? Most people would guess they are not. Until just a few hundred years ago, even scientists thought the Earth was still and fixed at the center of the universe. Now we know that idea is far from the truth.

Our planet Earth is orbiting the Sun at the enormous speed of 19 miles every second. On top of that, the Earth spins around its own axis once a day, and the Sun and its family of planets are moving around the Milky Way Galaxy. But the astronomers of 2,000 years ago had no way of knowing these things.

Greek philosophers

The Greek philosopher Plato (427–347 B.C.) taught that all heavenly bodies move at a constant speed along circular paths. His most famous pupil, Aristotle, lived from 384 to 322 B.C. and worked mainly in Athens. Aristotle was quite certain that the Earth must be the center of all things. The Sun, Moon, and planets, he believed, move around the Earth, following paths based on circles. Some years later, Aristarchos of Samos (about 310–230 B.C.) put forward his own idea that the Sun was at the center of the planetary system. But because there was not much observa-

Light takes 100,000 years to cross the Galaxy.

Light from the farthest galaxies takes about 10 billion years to reach us.

tional evidence to support the idea at the time, he did not succeed in persuading others that his theory was the correct one.

Claudius Ptolemaeus (Ptolemy), who worked in Alexandria, in what is now Egypt, detailed the Greek scholars' view of the Earth-centered universe in a book written between about 127 and 151. Nobody seriously challenged Ptolemy's model until the 16th century.

To the Greek philosophers, the circle was a perfect shape and the only one possible for an orbit. It never occurred to them that an orbit could be another curved shape. They had a problem, though, explaining the observed motion of the planets if they were supposed to travel evenly around the Earth in simple circles. To reproduce the uneven way the planets were seen to move through the sky, Ptolemy had to assume they move around small circles whose centers are in turn going around the Earth in circles. This sounds very complicated, and it is! However, the two ideas that orbits had to be circles and that the Earth was the fixed center of the universe were accepted as standard.

The Sun at the center

Nicholas Copernicus (1473–1543) established a new way of thinking. He believed, we now know correctly, that the Sun is at the center of the planetary system and that the Moon orbits around the Earth. Nevertheless, he had problems predicting exactly where the planets would be, because he kept the Greek idea of circular orbits.

The next breakthrough came when Johannes Kepler (1571–1630) realized that the orbits of the planets are ellipses (regular ovals), and not perfect circles.

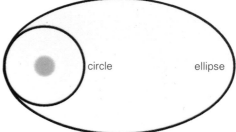

circle   ellipse

◁ The Earth-centered universe, favored in the Middle Ages, as shown in a 16th-century book. Ptolemy's model could not give accurate predictions for the positions of the planets.

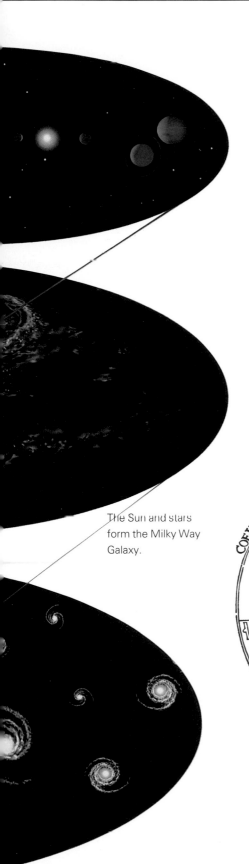

The Sun and stars form the Milky Way Galaxy.

All the galaxies make up the universe.

▷ The Copernican model of the universe, with the Sun (sol) at the center of the solar system. Copernicus published his ideas in 1543

The model developed by Copernicus, and improved by Kepler, eventually changed the way people thought about the universe and about science. Kepler brought in the idea of a force acting between the Sun and planets. Sir Isaac Newton showed how the motions of the planets could be explained by his theory of gravity.

Today we can say for certain that Earth is a planet, the third out from the Sun in a family of nine major planets. Many have moons orbiting them in turn, like miniature solar systems. Earth has just one Moon. The orbits of the planets are almost in one plane, as rings laid out on a table would be, so the solar system is shaped like a large, thin disk.

The astronomical unit

In round numbers, the average distance between the Earth and the Sun is about 93 million miles. Instead of writing millions of miles all the time, astronomers call this distance one astronomical unit (AU). The exact definition of the astronomical unit is 149,597,870 kilometers (92,955,730 miles). Jupiter is 5.2 AU from the Sun, Saturn is 9.54 AU, and Pluto is nearly 40 AU.

The nearest stars

The Greek philosophers thought they would see the star patterns alter over the course of a year if the Earth went around the Sun, just as you see the scenery around you change during a trip. They did not notice any change but that was because the stars are much farther away than they ever imagined. The distance to the nearest star (apart from the Sun) is 270,000 AU, which is 4.2 light-years. Astronomers know of about 2,000 stars within 50 light-years of the solar system. Nearly all of them are very faint.

The Milky Way and the Andromeda Galaxy

With the naked eye, you can see just a few thousand stars. The situation changes dramatically, though, if you use a telescope. Many faint stars come into view. In one particular band around the sky, you would notice there are vast numbers of fainter stars. This is the Milky Way.

KEY DISTANCES

Earth to Moon: 238,600 miles
Earth to Sun: 93 million miles
Sun to nearest star: 4.2 light-years
Sun to center of our Galaxy: 25,000 light-years
Sun to Andromeda Galaxy: 2 million light-years

◁ The great spiral galaxy, known as M31, in the constellation Andromeda. This is the closest spiral galaxy to the Milky Way (about 2 million light-years away), and it is similar to our Galaxy. There is a bulge of stars in the central region, or nucleus. There are two satellite galaxies as well.

There is a small, misty patch in the constellation Andromeda, which you can just see with the naked eye if you know where to look on a really dark night. Even a small telescope shows an oval-shaped blob, but photographs taken with larger telescopes reveal a magnificent spiral galaxy. Known as M31, or the Andromeda Galaxy, it is a giant family of billions of stars. Though tilted at an angle to us, it is not hard to see that this galaxy is shaped like a disk with a bright bulge in the middle. The disk has arms in a spiral pattern containing not only stars, but clouds of glowing gas and lanes of dark dust hiding the stars behind. The concentration of stars at the middle, making the bright bulge, is the part you can see with the naked eye.

There is every reason to believe that our Galaxy is similar to many other galaxies astronomers observe. It is probably like the Andromeda Galaxy, but not as large. Looking into space in the direction of the disk of our own Galaxy, we are bound to see many more stars than if we look above or below the disk, since that is where the vast majority of stars are concentrated. The Milky Way is our view of the disk of our own Galaxy seen from the inside.

Galaxies beyond

The Andromeda Galaxy is the only spiral galaxy, apart from our own Galaxy, you can see without a telescope, but there are vast numbers of fainter galaxies scattered through space. We can see them tilted at all different angles and get a good idea of what they are like. All of the stars in the universe are in the galaxies. There are no stars between the galaxies. Our Sun is just a star like the others and belongs to a galaxy. Looking out at distant galaxies beyond our own, it is easy to see their shapes. But we see our own Galaxy from within. All the stars visible in the night sky belong to our Galaxy.

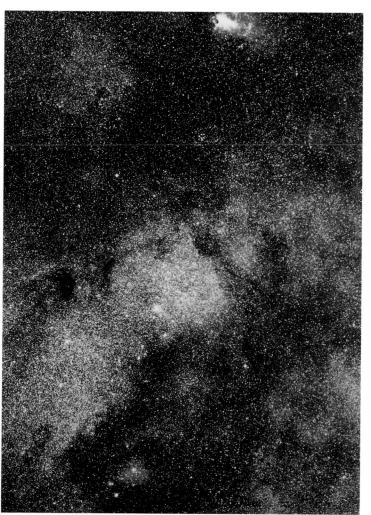

◁ Star clouds toward the center of the Milky Way. Millions and millions of stars crowd together, and the average distance between stars is much smaller than in the Sun's neighborhood.

△ Galaxies of the Fornax cluster, showing spiral and elliptical galaxies about 60 million light-years away.

EARTH AND SUN

The Earth spins daily on an axis that is tilted to its orbit around the Sun. Over the course of a year, the Earth makes a complete journey around the Sun and goes through the cycle of seasons with all the changes they bring. There are changes, too, in the stars that can be seen from night to night.

summer, Northern Hemisphere
winter, Southern Hemisphere

In the ancient world, people thought the Sun went around the Earth and nightfall came when the Sun's journey took it below the horizon. Paintings in Egyptian tombs show the Sun being pulled across the sky in a chariot. Now we know that as the Earth journeys around the Sun, it is also spinning on its own rotation axis once every 24 hours. The side of the Earth facing the Sun has day, and the other half has night.

Why, then, do we not have days and nights of equal length all year round? As it is, we find that the hours of daylight and darkness vary through the annual cycle of seasons. We get seasons because the Earth's rotation axis is not at right angles to its path around the Sun. If it were, there would be no seasons. But our rotation axis is tilted by about 23.5°.

The northern half of the Earth has summer when the North Pole is tilted toward the Sun. At midday on about March 20, the Sun is overhead at the equator. Then every day, until about June 21, the Sun is overhead at midday at more northerly places. On June 21, the Sun appears overhead at the Tropic of Cancer. This is midsummer's day in the Northern Hemisphere, when daylight lasts the longest. Its scientific name is the solstice. Places inside the Arctic Circle have some days when the Sun never sets at all. That is why countries in the far north are sometimes described as the Land of the Midnight Sun. After June 21, things gradually change back until the Sun is overhead at the equator again on about September 23.

While the Northern Hemisphere is having summer, the South Pole is tilted away from the Sun and the southern half of the Earth gets winter. The seasons in the south are exactly opposite those in the north. On December 21, the Sun is overhead at midday as far south as it ever gets, above the Tropic of Capricorn. This is the other solstice of the year, midsummer's day in the Southern Hemisphere.

On about March 21 and September 23, when the Sun is overhead at the equator, there are 12 hours of daylight and 12 hours of night all over the world. These days are called the equinoxes, which means "equal night."

Leap years

We say it takes a year for the Earth to go around the Sun, but it is not a year of exactly 365 days. The precise length of a year, say from one midwinter's day to the next, is 365.24219 days. If we did not put in a leap day every four years, the seasons would soon be out of step with the months, and that would be very inconvenient! To match the natural year even more accurately with our calendar, the rules say that years ending in 00, such as 1900, are not leap years unless they are exact multiples of 400. So the year 2000 will be a leap year, but 2100 will not. This system was introduced in 1582 to replace Julius Caesar's calendar.

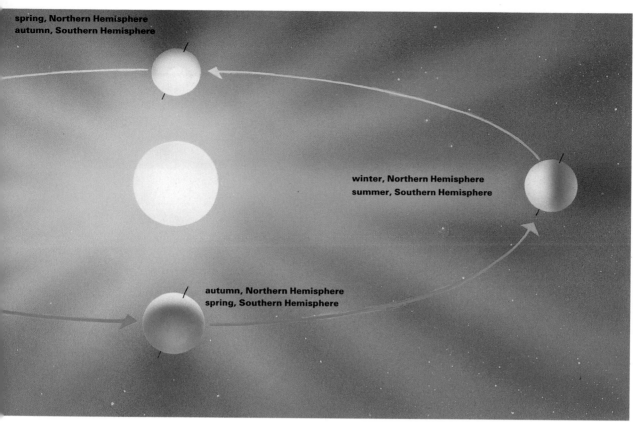

spring, Northern Hemisphere
autumn, Southern Hemisphere

winter, Northern Hemisphere
summer, Southern Hemisphere

autumn, Northern Hemisphere
spring, Southern Hemisphere

◁ The Earth's rotation axis does not make a right angle with the plane of its orbit around the Sun. Instead, it is tilted at 23.5° to the right angle of its orbit. If the Earth were not tilted, there would be no seasons and the day length would always be 12 hours everywhere. The tilt means that the regions pointing toward the Sun receive more hours of daylight and more solar heating than other parts of the Earth, and so these regions experience summer. On the equator the day length is always 12 hours and there are no major seasonal changes in the tropical regions.

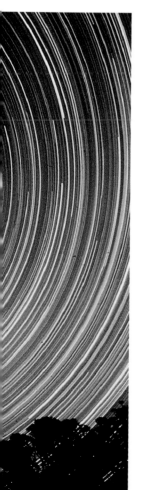

◁ A time-exposure photograph lasting all night shows stars wheeling around the south pole of the sky as the Earth rotates. Telescopes have to be guided accurately to keep in step with the stars; otherwise, star images would blur into trails on astronomical photographs as well.

The nightly parade of the stars

The rising of the Sun at dawn and its setting at nightfall are part of everyday experience that almost everyone takes for granted. But what of the stars? Here on the Earth's surface, we cannot see stars by day because the atmosphere scatters so much Sunlight, making the sky bright and blue.

Get to know the stars of the night sky and soon you will be aware that they, too, rise and set. After dark, the spinning of the Earth brings into view a constant succession of star patterns. New stars appear as they rise in the east, while others sink below the horizon in the west. According to where you are, some stars will never set. Instead, they sweep out circles around the pole of the sky. These are known as circumpolar stars or constellations.

You can prove to yourself that the stars are moving across the night sky. You will not detect their movement just by staring at them for a few minutes. The best way is to choose a bright star you can identify easily. See how it lines up with something like a rooftop or tree from a particular viewing spot. Then look again after one hour and after two. Its change in position will be obvious after such a length of time.

Changing seasons, changing sky

Though you can only see half the sky at any one time, as we have already discovered, you can see more and different stars as the Earth spins and they rise above the horizon. But not only does the sky change hour by hour, but what you can see varies from season to season as well. To understand why, we need to think about the Earth on its journey around the Sun.

Imagine it is midnight and you are looking out at the stars. You are of course on the side of the Earth facing away from the Sun. Now we wind the clock forward quickly to six months in the future. The Earth has completed half its journey around the Sun and its night side is pointing toward the direction in space directly opposite where you were looking six months ago. Your midnight observations now reveal a completely different part of the starry scenery. So, at different times of the year, the night side of the Earth points out toward different parts of the starry sky.

If you follow the stars over several months, you will notice how the patterns of stars visible at a particular time of night

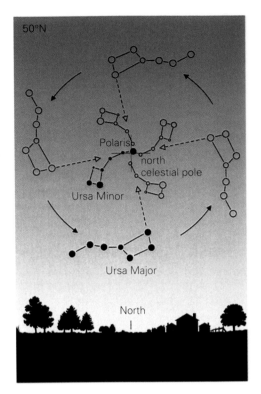

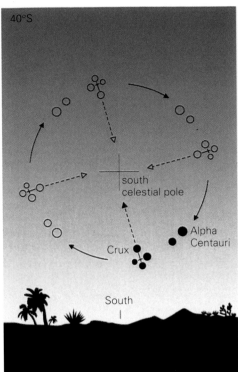

MEASURING IN ANGLES

The sizes of objects and the separations between them as they appear in our sky are measured in degrees. Across the heavens from the east to the west is 180° of angle. In terms of angles, the Sun and the Moon are both about 0.5° across. Almost all the objects studied by astronomers are so far away that their angular sizes have to be given in seconds of arc. There are 3,600 seconds of arc in a degree. A typical optical telescope used by a professional astronomer can see detail down to about half a second of arc under excellent observing conditions.

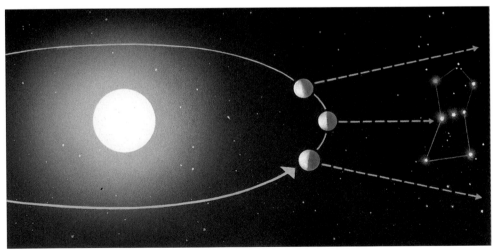

△ As the Earth orbits the Sun, the view of the night sky from a fixed location slowly changes.

rises at exactly the same time according to the sidereal clock.

For the Sun to get back to its same position in the sky, it takes a little longer than the other stars—a full 24 hours. You might think that in one day the Earth turns exactly once—through 360°. In fact, it has to turn through a little bit extra, about 361° altogether, before the Sun gets back to the same place in the sky. This is because the Earth has also moved a bit farther along its curved orbit around the Sun in the course of the day.

Stars all around: the celestial sphere

Even the nearest stars are much farther away than the Sun, Moon, and planets, so the patterns they appear to make in the sky do not change perceptibly as the Earth travels around the Sun. The starry sky is like distant permanent "scenery" surrounding the Earth and the whole solar system. If you were to travel a little way out into space away from the Earth, you would be able to see stars set in a dark black sky in every direction all around.

In the days before astronomers discovered that stars are scattered through our Galaxy, all at different distances, many ancient peoples, such as the Greeks,

change with the seasons. Try choosing a bright star that you can pick out easily from the constellation patterns each time you look. On clear nights, see where it is from your regular observing post, checking at the same time on each occasion. Try to follow the same star for a few weeks.

Sidereal time

At midnight in January the stars you can see are completely different from those on view at midnight in July. But there is not a sudden jump. Each night there is a small but noticeable change: any particular star rises about four minutes earlier than it did the previous night.

As the Earth spins in space, the length of time it takes a star to get back to the same position in the sky is only 23 hours, 56 minutes and 4 seconds. This short day is called the sidereal day. *Sidereal* means "pertaining to the stars." In an observatory the clocks run on sidereal time and lose four minutes each day compared with normal clocks. Astronomers like to use sidereal time because each star always believed the Earth was at the center of the universe. They imagined that the stars were fixed on some gigantic sphere surrounding the solar system, because that is exactly how things look. It is such a useful idea when we just want to talk about the positions of stars in the sky and not their actual locations in space that astronomers still use the concept of the "celestial sphere."

Although the celestial sphere surrounds the Earth completely, any observer on the Earth's surface sees just one half of the celestial sphere at a time because the ground gets in the way. Only astronauts out in space can see the stars below them as well as above.

The Earth spins around an axis between the North Pole and the South Pole. The equator is an imaginary circle around the earth halfway between the two poles. The celestial sphere looks to us as if it spins around

its north and south poles. The celestial equator is a circle around the sky exactly halfway between the poles. The poles and equator of the sky are exactly over the poles and equator of the Earth.

Right ascension and declination

We can say where any place is on the Earth by giving its latitude and longitude. Latitude is measured in degrees north and south of the equator. Longitude is measured in degrees east and west of a circle chosen to go through the North and South Poles and a marker in Greenwich, London.

To say where a star is located in the sky, astronomers use a grid on the celestial sphere similar to latitude and longitude on the Earth. The equivalent of latitude is declination measured in degrees from the celestial equator.

The equivalent of longitude is right ascension. It is measured from the point where the Sun is on the celestial equator at the March equinox. Because of the way the wheeling sky marks out time, right ascension is often measured in hours, minutes, and seconds rather than degrees. One hour is equivalent to 15 degrees.

The ecliptic

Because the Earth's axis is tilted, the Sun's path in the sky does not follow the celestial equator. The circle it follows is tilted from the equator by 23.5°, and it is called the ecliptic.

north celestial pole
declination +90°

Star

declination line

Right ascension line

ecliptic

line of latitude

North Pole

Earth

23.5°

line of longitude

♈

South Pole

First Point of Aries

celestial equator
declination 0°

south celestial pole
declination −90°

▷ The concept of the celestial sphere helps astronomers to describe the positions of stars in the sky. Right ascension is measured from the First Point of Aries, where the Sun crosses the equator in March.

EARTH AND MOON

Each month the Moon orbits the Earth and you can see a complete cycle of the Moon's phases. The Moon's shape changes from night to night, from crescent through to Full Moon, then back to a crescent again. The Moon also causes the ocean tides twice a day, and these also vary through a monthly cycle.

The Moon is Earth's companion in space. Once a month the Moon makes a complete journey around the Earth. It shines only by reflecting light from the Sun, so at any time the half of the Moon facing the Sun is lit up and the other half is in darkness. How much of its sunlit hemisphere is visible to us depends on the Moon's position in its orbit around the Earth. As the Moon moves in its orbit, the shape in which it appears to us is gradually changing all the time. The different shapes are called phases. The whole cycle of phases repeats itself after 29.53 days.

To understand why we see phases, first think of a New Moon. This is when the Moon is between the Earth and the Sun. The whole of its bright side is turned away from us toward the Sun and we cannot see the Moon. Now think of a Full Moon.

When the Moon is on the side of the Earth opposite the Sun, we see the whole of its illuminated half.

Sometimes, when the Moon is a very thin crescent, the rest of its disk is faintly visible. People say it is "the new Moon in the old Moon's arms." It is really sunlight bouncing first off the Earth and then off the Moon, so it is also known as "earthshine."

When can you see the Moon?

People often imagine the Moon is only up at night, but it is often possible to see the Moon faintly during the day if the sky is clear. The time of moonrise gets later from one day to the next. Just after New Moon the Moon rises after the Sun. By First Quarter, a week later, it is rising at midday, and at Full Moon it rises at sunset.

Harvest Moon

In autumn in the Northern Hemisphere, the Full Moon nearest the equinox on September 23 is popularly known as the Harvest Moon. For a few nights the Moon rises at almost the same time each evening, just as the Sun sets. In days gone by, farmers took advantage of this moonlight to complete their harvesting and so it became known as the Harvest Moon. The Moon looks bigger when it is near the horizon, but that is just an optical illusion.

Tides

Tides are familiar to anyone who lives near or visits the coast. Twice a day, the level of the ocean rises and falls, in some places by a very considerable amount.

▷ Halfway between New Moon and Full Moon, we can see just half of the Moon's bright side. The shape of the Moon is a semicircle. Because we are seeing just one-quarter of all the Moon's surface at this time, the half-Moon phases are traditionally known as First Quarter and Last Quarter. The in-between phases are described as crescent or gibbous, according to whether less (crescent) or more (gibbous) than a half-Moon can be seen.

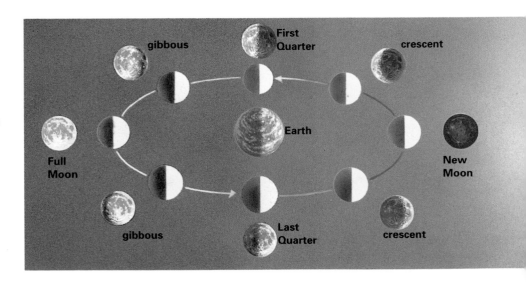

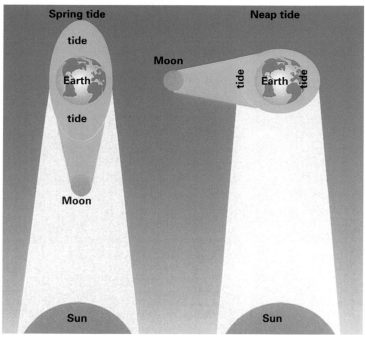

Each day, the tides come about 50 minutes later. What causes the ocean to slosh about? The Moon is responsible.

The Moon keeps orbiting the Earth because the two bodies are held together by the pull of gravity each has on the other. All the time the Earth is tugging at the Moon, and the Moon pulls at the Earth.

Because the oceans are liquid and can flow, they are easily distorted by the Moon's gravity into a lemon shape. The rocky ball of the Earth stays in the middle. The result is a bulge of water on the side of the Earth toward the Moon, and another bulge on the opposite side. As the solid Earth spins, the coasts experience high and low tides twice in a period of 24 hours and 50 minutes when they pass through the water humps. This time is longer than 24 hours because the Moon itself is moving along its orbit as well. In bays and river estuaries the rise and fall of the tides can be greater than in other places as the seawater is funneled into narrow channels.

Spring and neap tides

As well as the Moon, the Sun contributes to the height of the tides. When the Sun and Moon are lined up so that their gravitational pulls are in the same direction and reinforce each other, there are higher tides than average, called spring tides. This happens with the New Moon and Full Moon. When the Sun and Moon are pulling at right angles, they partly cancel out each other's effect and the tides are lower than average. These are neap tides, and they occur when the Moon's phase is

△ The crescent Moon showing earthshine (far left) and Earth and Moon imaged by the *Galileo* spacecraft (left).

△ Tides are raised by the pull of the Sun and the Moon on the oceans. The pull of the Moon on the Earth is strongest on the side closest to the Moon. The center of the Earth is pulled a little less and its far side even more weakly. In effect, the Moon is trying to stretch the Earth out.

First Quarter or Last Quarter. In practice, high and low tides happen a bit later than might be expected from the Moon's position. This is because of the drag of the water over the surface of the Earth.

Even the solid Earth is stretched slightly by the tidal forces of the Moon. The change amounts to about 12 inches. In turn, the Earth pulls the Moon out of shape by about 16 inches.

In principle, the energy of the movement of tides could be captured by building dams across large estuaries, and making the flow of seawater turn electrical generators. To give some idea of the amount of energy in the ocean tides, 6 percent of all the electricity used in Great Britain could be generated by a dam across the estuary of the River Severn in southern England. In practice, there are major environmental and engineering problems to solve before tidal energy could be harnessed in an acceptable way.

Longer days

Tides cause friction between the Earth's surface and the oceans, slowing down the Earth's rotation. Our days are slowly becoming longer and longer, by about two-thousandths of a second per century. Evidence of this slowdown comes from certain corals that grow in a way that leaves a fine ridge in the coral each day. The growth varies over the year, causing a band like the growth ring in a tree. By studying fossil corals 400 million years old, geologists have found that at that time the year had over 400 days of 22 hours each. Even more ancient life-forms have left fossils showing that the day lasted only 10 hours some 2 billion years ago.

Far in the future, a day will last for a whole month. What do you think the sky would look like? The Moon would always be in the same place, because the Earth's spin would be exactly in step with the Moon's motion along its orbit.

Already, because of the tidal forces between the Earth and Moon, the Moon always keeps the same face toward the Earth, apart from a small amount of wobble. The Moon is also speeding up in its orbit. As a result, it is steadily moving farther away from the Earth at a rate of 1.5 inches per year.

ECLIPSES OF THE SUN AND MOON

The Sun, the Moon, and the Earth carry out a complex "dance" in space. Sometimes all three end up almost in a straight line. That is when the fun of hide-and-seek starts, and you may be able to see an eclipse.

The Sun is 400 times bigger than the Moon, but it is also roughly 400 times farther away. By this accident of nature, the Sun and Moon as we see them in the sky appear to be almost the same size. As a consequence, it is possible for the Moon just to cover the Sun if its path happens to cross precisely between the Sun and the Earth. When the Earth, Moon, and Sun are in a completely straight line, there is a total eclipse of the Sun.

Every month, at New Moon, the Moon passes between the Sun and Earth in space, but the exact lineup we need for an eclipse is not so common. Even when it does occur, there is only a limited area on the Earth's surface from which the total eclipse is visible. This area, called the eclipse track, is long and narrow. It is never more than 167 miles wide but can be thousands of miles long. The longest time a total eclipse of the Sun can last at one place is just under eight minutes. In reality, most are shorter than this.

The sight of a total solar eclipse is so dramatic that some people travel halfway around the world to see one. The eclipse starts when the Moon first edges over the Sun's disk. It looks almost as if a piece is being eaten out of the Sun. During this time, the eclipse is only partial. Gradually, more and more of the Sun is covered until only a narrow crescent remains.

Outside the region where the view of the eclipse is total, people in a much larger area of the Earth's surface see a partial eclipse. Sometimes a partial eclipse takes

▽ A beautiful solar eclipse, March 18, 1988. As a solar eclipse reaches totality the sky darkens, and the air temperature falls. Flowers may close their petals, and birds start to roost. The faint outer part of the Sun, its corona, can be seen around the dark circle of the Moon because it is no longer swamped by the intense light of the Sun's yellow disk. Pink, flamelike prominences shooting out of the Sun's surface can often be seen projecting around the dark circle of the Moon. Bright stars can be seen in the dark sky.

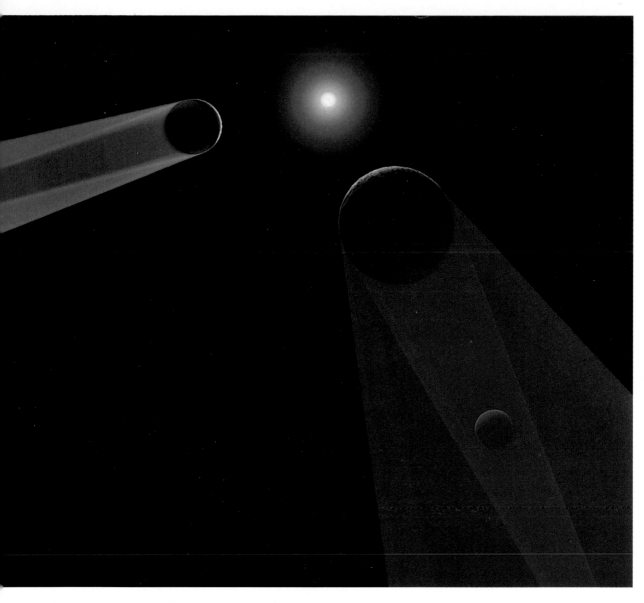

◁ If the Moon's shadow falls on the Earth (left), a solar eclipse occurs; when the Moon goes through the Earth's shadow, there is a lunar eclipse (right).

ECLIPSES OF THE SUN 1995–2005		
Date	Location	Type and duration
October 24, 1995	Iran, India, Pacific Ocean	Total, 2 m 5 s
March 9, 1997	Arctic, Russia	Total, 2 m 50 s
February 26, 1998	Pacific and Atlantic Oceans	Total, 3 m 56 s
August 22, 1998	Indian and Pacific Oceans	Annular
February 16, 1999	Indian and Pacific Oceans, Australia	Annular
August 11, 1999	Atlantic Ocean, Europe, India	Total, 2 m 23 s
June 21, 2001	Central Africa	Total, 4 m 57 s
December 14, 2001	Pacific Ocean, Central America	Annular
June 10, 2002	Pacific Ocean	Annular
December 4, 2002	Southern Africa, W. Australia	Total, 2 m 4 s
May 31, 2003	Arctic	Annular
November 23, 2003	Antarctic	Total, 1 m 57 s
April 8, 2005	Panama, Colombia, Venezuela	Annular/Total, 42 s
October 3, 2005	Portugal, Spain, N. and E. Africa	Annular

place that is not seen as total from anywhere on the Earth.

In some solar eclipses, the Moon passes directly in front of the Sun, but a bright ring of the Sun remains visible around the dark disk of the Moon. The distances between the Earth and the Sun and between the Earth and the Moon vary by small amounts. If the Moon is a bit farther away than average, it will look smaller; if the Earth is a bit nearer to the Sun than average, it will look bigger.

If this happens to be the situation when there should be a total eclipse, the Moon is not big enough in our sky to cover the whole of the Sun. Instead, we get an annular eclipse. (*Annular* means "shaped like a ring.") In an annular eclipse, the sky stays bright and you cannot see the Sun's corona.

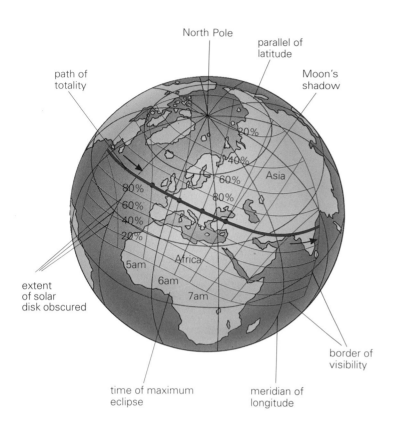

North Pole

parallel of latitude

path of totality

Moon's shadow

20%

40%

Asia

60%

80%

80%

60%

40%

20%

extent of solar disk obscured

5am

Africa

6am

7am

border of visibility

time of maximum eclipse

meridian of longitude

▷ A lunar eclipse, in which the Moon appears copper-red in color. The Moon never goes completely dark in a lunar eclipse because some sunlight is deflected through the Earth's atmosphere and dimly illuminates the Moon.

◁ The path of the eclipse that will take place on August 11, 1999, showing the regions where a total eclipse and a partial eclipse will be visible. The total eclipse will last only 2.5 minutes, occurring during the middle of the day over most of Europe.

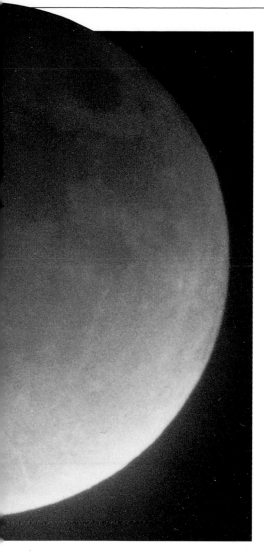

Date	Duration
TOTAL ECLIPSES OF THE MOON 1996–2004	
April 4, 1996	1 h 24 m
September 27, 1996	1 h 12 m
September 16, 1997	1 h 6 m
January 21, 2000	1 h 16m
July 16, 2000	1 h
January 9, 2001	30 m
May 16, 2003	26 m
November 9, 2003	11 m
May 4, 2004	38 m
October 28, 2004	40 m

Lunar eclipses

The Earth casts a long shadow through space, cutting out the light of the Sun. When the Moon goes into the Earth's shadow, a lunar eclipse takes place. If you were on the Moon during a lunar eclipse, you would see the Earth going across in front of the Sun. Often, the Moon can still be seen dimly, shining with a reddish tinge. Though it is in shadow, the Moon is lit by a small amount of red sunlight scattered in its direction from the Earth's atmosphere. Totality can last for up to 1 hour 44 minutes. Unlike solar eclipses, lunar eclipses can be seen from anywhere on the Earth where the Moon is in the sky.

Eclipses long ago

Eclipses of the Sun and Moon greatly interested people in ancient times. The philosophers in ancient Greece were convinced that the Earth is a sphere because they noticed that the edge of the Earth's shadow on the Moon is always circular. Furthermore, they came to the conclusion that the Earth is about three times bigger than the Moon, simply by observing how long the eclipse took. (The correct value is 3.66 times larger.)

Evidence from archaeology suggests that many ancient civilizations tried to predict eclipses. Observations at Stonehenge in southern England may have enabled people in the late Stone Age, 4,000 years ago, to predict some eclipses. Certainly they could work out when midsummer and midwinter would occur.

In Central America 1,000 years ago, the Mayan astronomers were able to foretell eclipses by making a long series of observations and looking for repeated patterns. Almost the same eclipses repeat every 54 years, 34 days.

How often can we see eclipses?

Though the Moon travels around the Earth once a month, eclipses cannot take place each month because of the way the Moon's orbit is tilted relative to the Earth's. At most, there can be seven eclipses in any one year, of which two or three must be eclipses of the Moon.

Solar eclipses only take place at New Moon, when the Moon is directly between the Earth and the Sun. Lunar eclipses occur only at Full Moon, when the Earth comes between the Moon and the Sun.

You can expect to see 40 lunar eclipses in your lifetime (assuming clear skies). Solar eclipses are much more difficult to see because the eclipse tracks are so narrow. Most people are unlikely to see a total solar eclipse without planning a trip to a future eclipse track.

◁ These images show the progress of a total eclipse of the Sun. The first three show the partial phases. In the fourth image the Sun's outer atmosphere, the corona, is briefly visible. The fifth photograph shows the Diamond Ring effect, when sunlight shines through a valley at the edge of the Moon.

CONSTELLATIONS

In ancient times our ancestors divided the starry skies into clear patterns named constellations. Several ancient cultures related the patterns in the stars to their traditional gods and myths. Modern star charts still use these mythical sky pictures as a basis for mapping the skies.

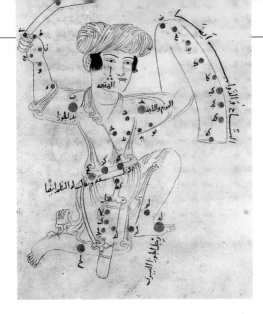

△ The constellation Orion (the Hunter) pictured by Al-Sûfi, a 10th-century Islamic astronomer. Three stars make Orion's belt.

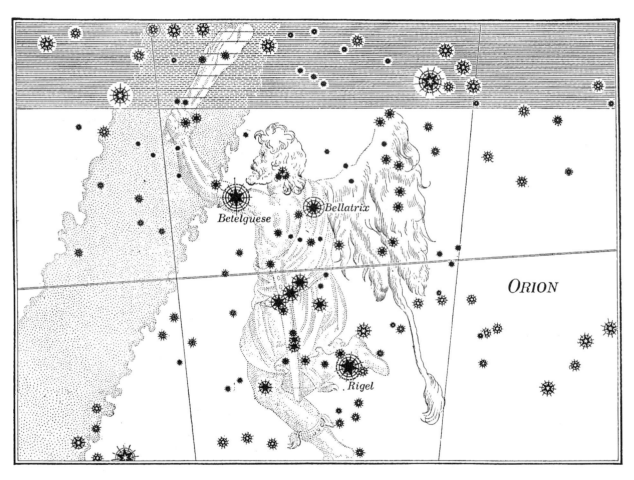

◁ An 18th-century map of the Orion constellation, copied from Albrecht Dürer's engraving.

Anyone looking at the night sky soon begins to link the brighter stars into simple patterns, such as squares, triangles, lines, crosses, and arcs. It does not take much imagination to begin to see the outlines of some objects: a flying bird, a crown, or the long curled tail of a scorpion.

Extend the idea a bit further, and what simpler way could there be of identifying all the brighter stars in the sky than to talk about the patterns that they make? In this way, no doubt, the concept of constellations was established and developed.

From the earliest records we have, it is clear that different civilizations have always given names to the more obvious star patterns. All astronomers, including professionals, still use them today.

The constellations are a useful way of helping to identify the positions of stars on the celestial sphere, but stars that appear to be close to each other in the sky may be separated by vast distances in space and have no actual connection between them. The stars are scattered all through our Galaxy. The patterns in which we see them are for the most part just chance. For example, the three brightest stars in the Southern Cross are actually at distances from the earth of 360 light-years, 420 light-years, and 88 light-years.

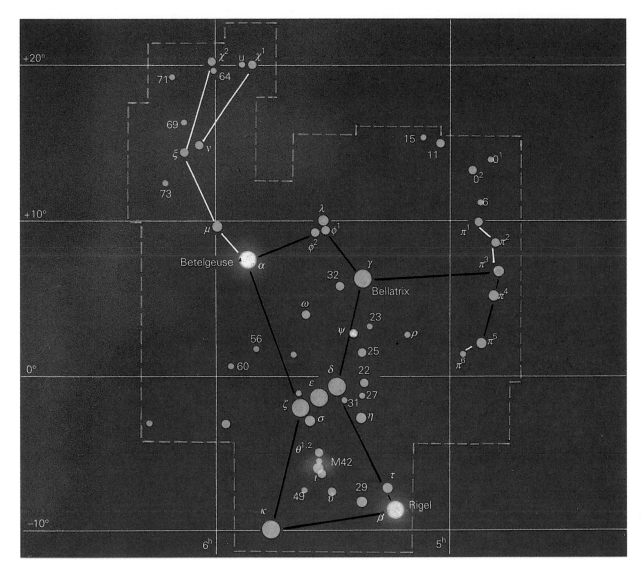

Modern constellations

As more and more faint stars were discovered, particularly after the telescope was invented, astronomers were not sure how to relate them to the bright star patterns. Certain stars might be put in one constellation in a particular star atlas and in a different constellation by another mapmaker. In 1930 astronomers worldwide agreed to divide the whole of the sky into 88 regions, based on traditional constellations. These regions are what astronomers now mean when they talk about constellations, rather than the patterns made by the brightest stars.

Of the 88 constellations now recognized officially, 48 were listed by Ptolemy in the 2nd century and many were probably in use a long time before then. Most of their names are linked to the stories of Greek mythology. Hercules, Orion, Perseus,

Andromeda, and Pegasus are just a few examples. The official constellation names are in Latin.

The ancient Greek philosophers and their predecessors lived around the Mediterranean and the region we call the Middle East. From there they could never see parts of the southern sky. Not until explorers started to travel in the south did European astronomers give names to the constellations in this part of the sky, so they have all been added since about 1600. Many were invented by the German Johann Bayer (1572–1625) and an additional 14 by Nicholas de Lacaille (1713–62), a French astronomer. In contrast to the ancient mythology of the north, Lacaille favored items of technical equipment and introduced the Chemical Furnace (Fornax), the Telescope (Telescopium), the Microscope (Microscopium), and the Clock (Horologium), among other curiosities.

Learning the constellations

If you do not know any constellations at all, it can be difficult getting started, but once you have identified one or two, hopping to others nearby soon helps build up your knowledge. To begin with, you might need help from someone who can point out some constellations. A simple star chart for the date and the time when you want to observe is a great help. As we have seen, the sky changes markedly both through the night and from one week to the next. This can be very confusing when you first start as a constellation hunter. Furthermore, what you can see also depends on your latitude. The simple charts you will find at the end of this book (pages 150–153) should help you find the more prominent constellations visible from your own location at different times of the year.

A number of star patterns within constellations have their own traditional names, even though they are not official constellations. These named star patterns are termed asterisms. Perhaps the best known is the Big Dipper, in the constellation of the Great Bear (Ursa Major). Another example is the Sickle in the Lion (Leo).

The zodiac

The names of 12 particular constellations are probably better known than most of the others. These are the ones forming the zodiac. The word *zodiac* means "belt of living things" because almost all of these constellations are named after animals or mythical beings. Libra (the Scales), however, is the one zodiacal constellation that does not fall into that category.

The zodiac gained its importance in astronomy because the yearly path of the Sun through the stars, known as the ecliptic, lies within this band. The paths of the Moon and planets through our sky are also in the zodiac.

During the day, you cannot see the stars, but they are still there of course. If you could see the daytime stars, you would be able to see directly the constellation in which the Sun is located. This is actually

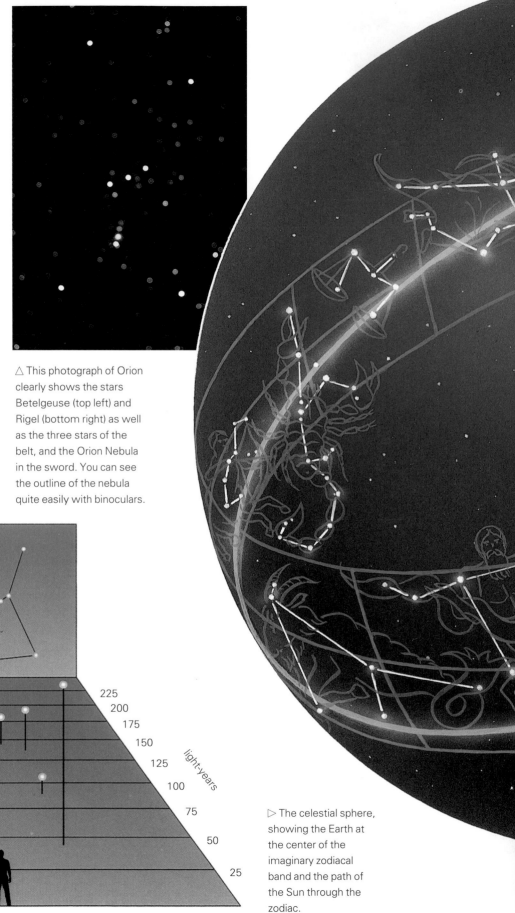

△ This photograph of Orion clearly shows the stars Betelgeuse (top left) and Rigel (bottom right) as well as the three stars of the belt, and the Orion Nebula in the sword. You can see the outline of the nebula quite easily with binoculars.

▷ The stars of Orion are really all at different distances from the Earth. Over millions of years the appearance of constellations will change as the individual stars move relative to each other.

225
200
175
150
125
100
75
50
25

light-years

▷ The celestial sphere, showing the Earth at the center of the imaginary zodiacal band and the path of the Sun through the zodiac.

possible during the few minutes of a total solar eclipse.

For the same reason that the constellations visible at night gradually change over the course of a year, the Sun appears to make one complete trip around the sky during a year. The modern constellations of the zodiac are not of equal size, so the Sun spends different lengths of time in each. Furthermore, the ecliptic is actually slipping slowly around the sky, so that it now passes through a thirteenth constellation, Ophiuchus.

Astronomy and astrology

In contrast to the true astronomical situation, astrologers choose to divide the zodiac band into 12 equal parts, called "houses," which are given the same names as the 12 traditional zodiac constellations. The astrological houses no longer correspond with the positions of the real constellations. Your astrological "sign" is determined by which house the Sun lies in at the date of your birth. The Sun is generally not in the real constellation of that name on your birthday.

Astrology probably arose when people first realized that the Sun and the Moon have a direct effect on tides, the weather, and the seasons. From this it must have seemed natural to believe that the planets might also have some effects. There are many references to the influence of the stars in the works of great poets such as Virgil (70–19 B.C.), Chaucer (1340–1400), and Shakespeare (1564–1616).

In 1226 the Mongol conqueror Genghis Khan stopped his wars when astrologers warned him that Jupiter was about to overtake Saturn. This seemed a bad omen for the campaign. The great astronomer Johannes Kepler started his career by casting horoscopes, which attempt to predict the future, for his friends. His later work, which showed that there are physical forces behind planetary motion, removed claims that the folklore of astrology had any scientific basis.

Today, most astronomers do not take astrology seriously. Centuries ago, however, many useful astronomical observations were made because of the belief that the stars could affect human affairs. Today, there is no connection between astronomy and astrology.

ORBITS IN THE SOLAR SYSTEM

The Sun has a family of nine major planets, along with many smaller bodies, such as asteroids and comets. Gravity holds the family together as each planet moves in an elliptical orbit around the Sun. The exploration of the planets by spacecraft is a major scientific achievement of the 20th century.

△ Sir Isaac Newton (1642–1727)

The ancient Greek astronomers had a very complicated idea of what the solar system looked like. This was because they thought that the Earth was at the center and that the orbits must be perfect circles. Even so, the Greeks came up with ways of predicting the future positions of planets that worked quite well. Nicholas Copernicus correctly placed the Sun at the center of the solar system, but he then had trouble predicting the positions of the planets because he held to the idea of orbits. Johannes Kepler was able to improve Copernicus's picture. Kepler found three rules to explain planetary motion, which are as follows:

1 Each planet travels around an ellipse, not a circle. The Sun is not at the center of this ellipse but off to one side, at a place called the focus. The more squashed the ellipse, the farther off-center the focus. This means that the distance between a planet and the Sun varies as the planet travels around its orbit.

2 A planet moves fastest when it is closest to the Sun and slowest when it is farthest away.

3 The time it takes a planet to orbit the Sun depends on its distance from the Sun. Kepler found there was a mathematical rule linking orbital period and distance from the Sun.

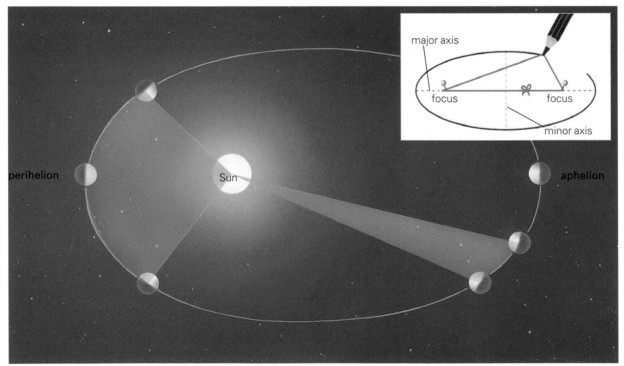

◁ You can draw an ellipse with a pencil, some string, and two straight pins. Keep the string tight and the pencil will trace an ellipse as you go around the loop.

◁ Illustrations of Kepler's first and second laws. Every planet travels on an elliptical orbit with the Sun at one focus. The speed of a planet varies, so that an imaginary line joining the planet to the Sun sweeps out equal areas in equal intervals of time. Perihelion is the point of closest approach to the Sun, and aphelion is the most distant part of the orbit.

Kepler thought some force from the Sun must drive the planets around. Whatever the force was, it must get weaker farther from the Sun. With Kepler's laws of planetary motion, astronomers could predict the positions of the planets about 10 times more accurately than before. However, Kepler could not say exactly what force held the planets in their orbits.

Isaac Newton and gravity

Sir Isaac Newton was the genius who explained motion in the heavens through the law of gravity. There is a popular story that Newton got his ideas in a flash while sitting under an apple tree and watching an apple fall, under the force of gravity. In fact, his theory was arrived at as the result of careful research over many years. He studied moving objects to try to understand how they behaved.

Newton realized that a force is constantly pulling at the planets to keep them from flying off into space. This led

him to devise his law of gravitation. According to this law, all material objects are attracted to each other. The more massive the objects, the stronger the pull between them, but the strength of the attraction decreases as they get farther apart.

Newton began to work on a theory of gravity because he thought the force that pulled a falling apple to the ground might also be the force that held the Moon in its orbit around the Earth. Three hundred years ago this seemed a crazy idea. Most educated people believed that the physical laws applying to the earth would not work for motions in the heavens. Newton's gravitational theory was the first scientific law that could be used to explain observations on the earth and in the heavens. It is true for everything that has mass: apples, people, moons, and planets.

For about 20 years Newton tried to describe the moon's orbit by using mathematics. His theory was a great success. The moon's motion had seemed very complex, but Newton found he could predict its

position with amazing accuracy.

Gravity and orbits

Newton's theory explained mathematically why planets and moons travel on elliptical orbits. His friend Edmond Halley (1656–1742) used this theory to predict the return of the comet now named after him.

The law of gravity also provides a way of measuring the masses of planets and their moons from their orbits. It works for pairs of stars that are orbiting each other, or for remote galaxies slowly moving within great galaxy clusters. Gravity holds together the stars of the Milky Way as one great galaxy.

Distances in the solar system

Newton's law of gravity allows astronomers to calculate distances to the planets, provided just one distance can be measured very accurately. To get one distance for starting the calculation, radar is used to measure the distance between the Earth and Venus or Mercury. In a radar measurement, the time it takes for a radio wave to travel to a planet and back is recorded with very great precision. We know that radio waves travel at the speed of light, so it is easy to determine how far they have gone.

Radio signals from spacecraft that have traveled to the planets also give us information on distances. By measuring the time it took radio signals to reach the Earth from the *Viking* spacecraft that landed on Mars in 1976, the distance to Mars could be worked out to within a few feet.

During the Apollo missions (1969–72), astronauts left reflectors on the Moon, like those used on bicycles and cars. By aiming a laser beam at the reflector and timing the round trip for light to go to the Moon and back, astronomers have measured the Earth–Moon distance to within less than three feet.

From the few distances that have been found with high precision, astronomers can calculate distances to all the planets and their moons.

With distances in the solar system known accurately, the masses of planets and their moons are calculated using the laws of gravity.

◁ A beam of light from a laser at the McDonald Observatory in Texas is used to measure the distance to the Moon. The time for a pulse of laser light to go to the Moon and back can be measured with great accuracy.

△ Satellites in a polar orbit travel over the North and South Poles. On an inclined orbit the satellite crosses the equator at an angle. The geosynchronous orbits are over the equator.

Satellite orbits

Satellites for observing the Earth are often put into orbit a few hundred miles up over the North and South Poles. The Earth rotates under this orbit, so each time the satellite goes around, it scans another north–south strip. Over time the entire globe is imaged.

Another common arrangement is to put a satellite in a circular orbit around the equator at an altitude of about 22,000 miles. In this path it takes exactly one day to orbit the Earth, so it is always over the same place on the Earth because it keeps pace with the Earth's rotation. These orbits are called geostationary or geosynchronous. They are used by TV and communications satellites. Small rocket engines on board the TV satellites are used to keep them in exactly the right place above the Earth.

Interplanetary travel

Space probes to the planets have to cover enormous distances. Just to reach Mars involves a journey of 280 million miles, even though Mars comes closer than 60 million miles every two years. To reach another planet, space scientists split the journey into three sections. First, the probe has to escape from Earth, which requires a minimum velocity of 25,000 miles per hour. The spacecraft is then positioned in an elliptical orbit around the Sun in which it coasts toward the target planet under the influence of the Sun's gravity alone. By choosing the right orbit, a spacecraft can be directed either toward the planets nearer the Sun than Earth, or those farther away. This saves an enormous amount of fuel but takes a lot of time—months in the case of Mars, and years for the outer planets.

As the probe draws level in orbit with the target planet, its on-board rockets are fired to move it into the final approach. When the probe reaches the planet, rockets are fired again to slow it down and place it into an orbit around the target.

For missions to the planets beyond Mars, space scientists sometimes use the gravity of a planet to give an extra boost to the spacecraft. The space probe is deliberately sent to a close encounter with another planet in order to fling it at a higher speed toward its target. The *Galileo* mission used this trick to send a probe to Jupiter: first the craft was sent to Venus for a gravity kick, and then it was flicked past the earth for another boost. Although this added several years to the traveling time, it meant a smaller launch rocket could be used.

▷ In the spectacular *Voyager 2* mission of the 1970s, space scientists guided *Voyager* past Jupiter, Saturn, Uranus, and Neptune in turn without using any booster rockets at all. In each case, the gravitational field of one planet was used to propel *Voyager* on to the next target.

forward motion

Moon

force of gravity

Earth

pull of string

◁ The tug of gravity holds a planet in orbit, just as the pull of a string holds a ball in an orbit as it is whirled around. Without gravity holding them in their orbits, the planets would simply fly off into deep space.

How Far Are the Stars?

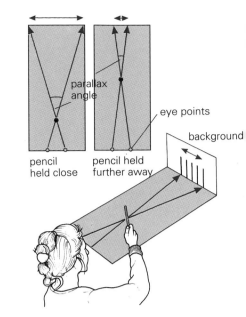

The nearest star (other than the Sun) is a quarter of a million times farther from Earth than the Sun, at a distance of more than four light-years. The farthest galaxies are billions of light-years from us.

How far away are the stars? Right away we can say that they are much farther away than the planets. If that were not the case, we would clearly see the patterns they make in the sky move around over the six months it takes for the Earth to travel 186 million miles from one side of the sun to the other. To understand why this is so, imagine you are in a car or train traveling through open country. As you move, the scenery changes. Trees near the road or railroad track go by quickly, while distant trees remain in view for a longer time.

In fact, over the space of six months, nearby stars do change their positions slightly compared with more remote ones, but the shift is so slight that it has to be measured accurately with the help of a telescope. You cannot see it with the naked eye alone. The effect is called parallax.

Distances to the nearest stars are found in exactly the same way, by measuring parallax. Imagine that the pencil in the diagram above is a nearby star, that your surroundings are more distant stars, and that your two eyes are two different viewing points in the Earth's orbit around the Sun. By taking measurements six months apart, astronomers get observations separated by 186 million miles. The shift in the star's position is still slight. If there were a star one light-year away, the total change in angle would be six seconds of arc (see page 28 to find out what a second of arc is). This is like seeing a small coin at a distance of 10 miles move about an inch. But even the nearest star is more than four light years away, so the change is even smaller than this.

△ To observe the parallax of a pencil held at arm's length, open and close each eye in turn. The pencil appears to jump in position relative to the background. The change in position is the parallax of the pencil and it can be measured as an angle. The farther away you hold the pencil, the smaller the parallax angle.

▽ Distances of the nearby stars can be found by measuring their change in position (or parallax) from opposite sides of the orbit of the Earth.

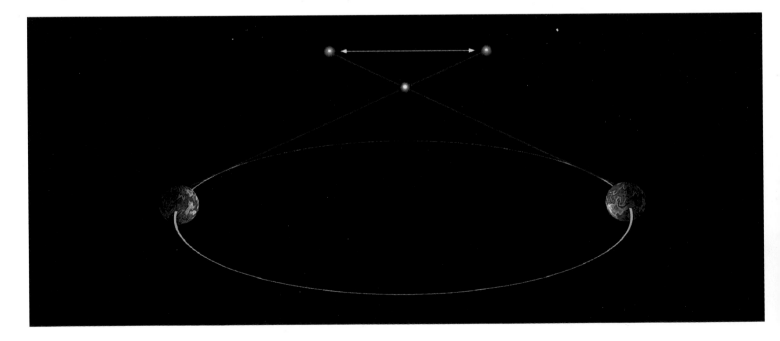

Using telescopes on the ground, star distances of up to about 60 light-years can be measured. The *Hipparcos* satellite, which orbited above the atmosphere between 1989 and 1993, collected vast amounts of data on star positions so that the distances of stars as far away as 200 light-years could be measured.

Stepping-stones to the stars

The vast majority of all known stars are too far away for the parallax technique to work. Astronomers have invented other ways to find distances. A common one is to measure how bright a star looks and compare that measurement with the total amount of light the star is actually emitting. The brightness we measure depends on how far the star is from us and on how much light it is sending out. If this second quantity can be determined from other properties of the star, the distance is calculated by comparing the star's real brightness with its apparent brightness.

The star brightnesses we measure depend mainly on two factors. The farther a star is from us, the fainter it will appear. If all the stars shone with the same light output, it would be easy to figure out their distances! However, the light output of stars—what astronomers call luminosity —covers an enormous range. Red stars are much less luminous than white ones. Big stars are more luminous than small ones.

Fortunately for astronomers, there are some stars that are instantly recognizable because they vary regularly in their brightness. These are known as regular variable stars. The time it takes for their light to rise and fall gives away their true brightness. To work out the distance of one of these stars, you only need to time how long it takes for the light to vary and how bright it appears to be on average.

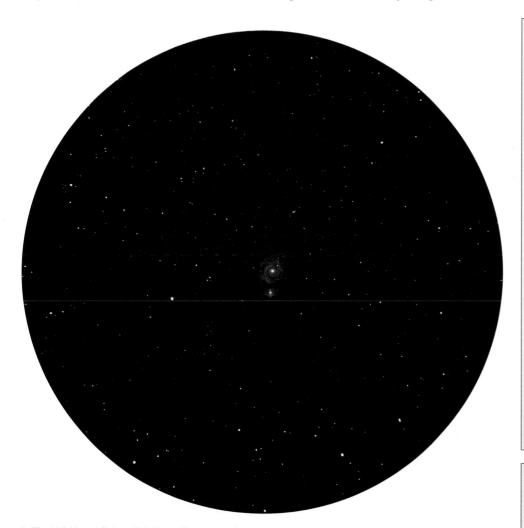

△ The Whirlpool Galaxy (M51) and its companion (NGC 5195). The spiral arms of galaxy M51 contain many young stars. Both galaxies are interacting through tidal forces, which have pulled one of the larger galaxy's spiral arms toward the smaller galaxy. These galaxies are about 30 million light-years away.

STAR MAGNITUDES

The brightness of stars is measured in magnitudes. On the scale of magnitudes, the lowest numbers are allotted to the brightest stars. The idea dates back to the Greek astronomer Hipparchus, who lived around 120 B.C. He divided the stars that could be seen with the naked eye into six groups according to brightness. He called the brightest stars "first magnitude" and the faintest "sixth magnitude." He had no telescope or other scientific instrument so his magnitude system involved guesswork.

In the 1850s the magnitude system was made scientific. It turned out that some of the brightest objects in the sky exceed first magnitude, so zero and negative numbers are used for them. Sirius, the brightest of all stars, as seen from the earth, is magnitude –1.46. A magnitude number greater than 6 means that an object is visible only through a telescope.

PARSECS

Professional astronomers use their own special distance unit, which they invented from measuring the parallax of stars. *Parsec* is a short form of *parallax-second*. One parsec (abbreviated as pc) is 3.26 light-years, a kiloparsec (kpc) is 3,260 light-years, and a megaparsec (Mpc) is 3.26 million light-years.

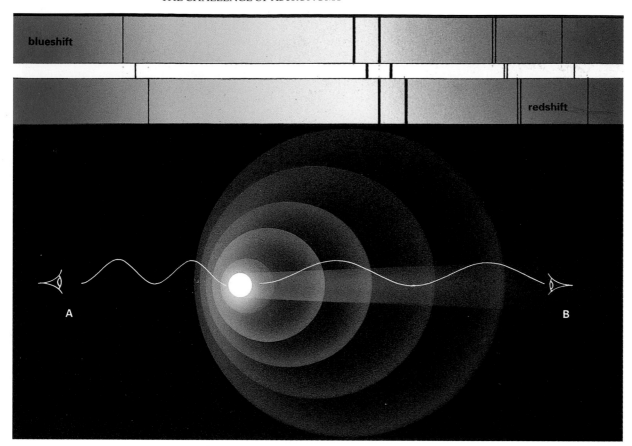

◁ Redshift and blueshift. The white spot represents a source of light moving toward the left. For observer **A** on the left, the light waves are bunched up and appear to have shorter wavelengths than if the light source were still. The pattern of dark lines in the spectrum is shifted toward the blue (upper spectrum). To observer **B** on the right, the light waves look stretched out and seem to have longer wavelengths. A redshift is seen in the spectrum (lower spectrum).

Variable stars extend the range

Measurements of regular variable stars put a big extension on our cosmic ladder to the universe. They expand our range right across our Milky Way Galaxy (100,000 light-years) and out to the nearest galaxies, such as the Andromeda Galaxy, which is about 2 million light-years away.

The type of variable stars called Cepheid variables are especially important, because they help astronomers work out distances beyond our own Galaxy. Cepheids are extremely luminous, so they can be picked out individually among the stars of nearby galaxies.

For more remote galaxies, astronomers can work out distances by seeing how bright exploding stars become, or by looking at the light we get from their largest star clusters. These things are much the same in one galaxy as in another, so any difference in brightness is a genuine effect of distance.

The redshifts of galaxies

In the 1920s the American astronomer Edwin Hubble (1889–1953) discovered a link between the distances of galaxies and their speeds through space. He photographed the spectra (light broken into the rainbow of colors) of many galaxies. Hubble used the 2.5-meter telescope at Mount Wilson, California, which was then the largest telescope in the world. For almost all the galaxies he studied, Hubble found that lines in the spectra were not in their usual positions. In many galaxies the spectral lines were closer to the red end of the spectrum than normal. The amount of shifting to the red varied from galaxy to galaxy.

The redshift is explained in the following way. When a galaxy moves away from us, its light waves are stretched out. The faster the galaxy goes, the more the waves are stretched. Longer light waves look redder to us. A similar thing happens to sound waves when a police car blasting its siren goes by at high speed. We hear the sound of the siren drop because the sound waves are stretched out.

For each galaxy, Hubble worked out the speed needed to cause the redshift he had measured. He found galaxies traveling away from us at enormous speeds, up to within a few percentage points of the

▽ Edwin Hubble's studies of galaxies were the first to suggest the huge size of the universe, when he showed that all of the other galaxies are millions of light-years from our Galaxy.

speed of light. (No galaxy can go faster than light.) Astronomers measure redshift by the amount the light waves are stretched. For example, if they are doubled in length, the redshift is 1.00. The more they are stretched, the greater the redshift.

What was really surprising was the discovery that almost all galaxies have redshifts. Only near our own Galaxy do we find a handful of galaxies with small blueshifts, indicating that they are moving toward us rather than away. The clusters of galaxies in the universe are speeding away from us and from each other.

Distances to galaxies

Hubble knew the distances to some of the nearer galaxies from observing their variable stars. He set about finding their speeds.

In 1929 he announced his results. He made a graph with velocity on one axis and distance on the other. Then he plotted the location of each galaxy on the graph. To his surprise, almost all the points on the graph made a diagonal straight line. Hubble had found that the galaxies are moving faster the farther away they are.

This discovery became known as Hubble's Law. It became the key to unlocking the secrets of the universe on the largest scale of distances. By finding the speed of a galaxy from its redshift, astronomers could find its distance from Hubble's Law.

How large is the universe?

Astronomers in the 1930s were astonished at how far away some galaxies were. We know today that the most distant detectable objects are around 17 billion light-years away. At the end of the cosmic distance ladder, there are many uncertainties. Nevertheless, light from the remotest parts of the universe set out on its journey billions and billions of years before our solar system even formed.

Our universe, as we can observe it, has its limit about 17 billion light-years away. At this limit the objects are traveling away from us at almost the speed of light. We do not know whether the universe is larger than this. Some astronomers believe that even the vast sphere we can view is only a speck in a gigantic universe far larger than anything that could ever be seen.

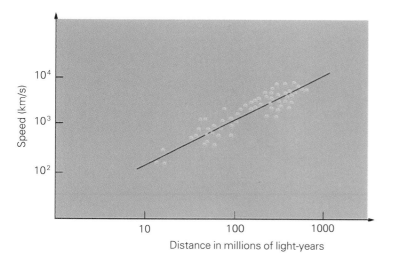

◁ The Hubble Law, showing how the distance of a galaxy is related to its measured speed. The greater the speed, the farther the galaxy is from us.

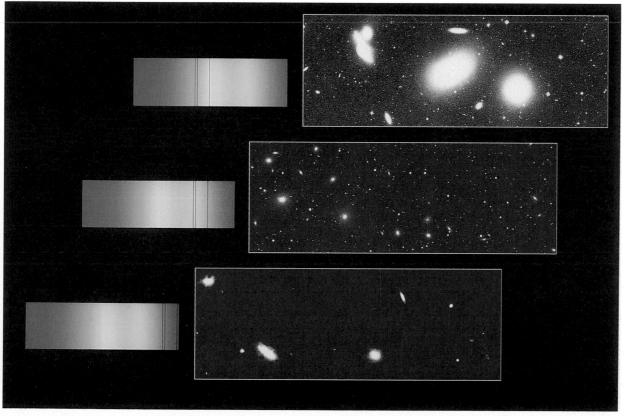

◁ Three galaxy clusters. Virgo (top) is about 65 million light-years away. Perseus (center) is 325 million light-years away. Galaxy cluster CL0939+4713 (bottom), photographed by the Hubble Space Telescope, is nearly 4 billion light-years away.

THE SOLAR SYSTEM

Where can you find volcanoes gushing with molten sulphur? Has there ever been water on Mars? Which planet captures comets and asteroids? The solar system is full of surprises: rocky planets and moons scarred with craters, giant planets made of gas, comets that appear unexpectedly, and stones that fall from the sky.

MERCURY

Mercury is closer to the Sun than any other planet, and completes each of its orbits of the Sun in only 88 days. It is the smallest planet, except for Pluto. The surface of this tiny world is hot enough to melt tin and lead. There is hardly any atmosphere, and the rocky terrain is covered with craters.

Mercury's orbit is closer to the Sun than the Earth's orbit. Because of this, Mercury is always quite close to the position of the Sun as seen in our sky. There are only a few days in each of its 88-day orbits when it is far enough from the Sun to be visible at all. At these times, it is low in the twilight sky. Unfortunately, Mercury is never seen in really dark conditions.

Mercury is best seen in the evening sky in spring or in the predawn sky in the autumn. You need to check in a yearly astronomy guide or the monthly astronomy column in a newspaper to catch the best days. Binoculars will give you a better chance of spotting Mercury, but you must not use them unless the Sun has set. Mercury will seem like a bright star. Once you have spotted it through binoculars, you may also be able to see it without using them.

Mercury's long days

Mercury spins on its own axis once in 59 Earth days. Long ago Mercury may have spun more rapidly, but the gravitational pull of the Sun has now slowed down its rotation.

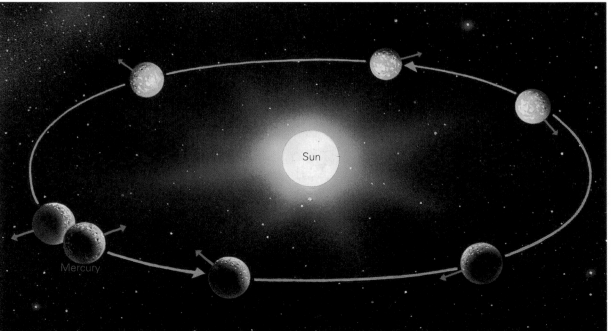

◁ Mercury takes 59 Earth days to turn once on its axis. Start where the arrow is pointing at the Sun. Now follow the orbit and see Mercury slowly turning as it journeys around its orbit. The arrow turns once when Mercury has moved two-thirds of the way around the Sun. It takes two full circuits of the orbit before the arrow again faces the Sun. This means that one day on Mercury lasts two years.

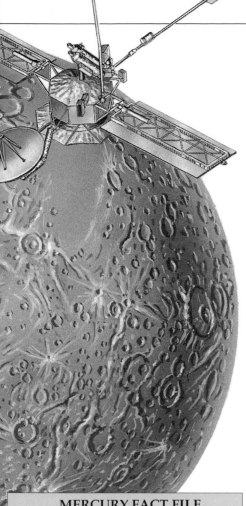

Mercury rotates three times for every two of its orbits of the Sun. During one rotation on its axis, Mercury completes two-thirds of its orbit around the Sun. So if you went to Mercury, you would find it took two "Mercury years," or 176 Earth days, from one sunrise to the next. You would also see the Sun change considerably in apparent size, because Mercury's orbit is quite elongated. During the very long "days," the temperature on Mercury's equator rises to 800°F, which is hot enough to melt tin, lead, and zinc.

Almost airless

Mercury has almost no atmosphere. Whatever gases may have been there in the past, everything has now been boiled away by the searing heat of the Sun. However, Mercury does manage to catch some of the wisps of hydrogen and helium gas that blow away from the Sun. Also, the baking hot rocks breathe out sodium atoms. So the very thin atmosphere is mainly sodium, with traces of helium and oxygen.

With no air and no clouds, the weather forecast on Mercury would be very simple: unbearably hot by day and, at the polar regions, freezing cold at night.

Surface like the moon's

It is impossible to see details on the surface of Mercury with a telescope because it is too far away. Over 10,000 images of the surface were made in 1974 and 1975 by the spacecraft *Mariner 10*. The best photographs show craters and surface cracks as small as 300 feet wide. *Mariner 10* photographed nearly half the planet. The surface is heavily cratered, as on the moon. One huge circular feature, the Caloris Basin, is 800 miles in diameter. This was probably made by a collision with a large asteroid. Just as on the Moon, there are small craters shaped like bowls, and larger ones with central peaks. A great many of the craters look as though they were caused by impacts with meteorites and asteroids.

Soon after the planets began to form, about 5 billion years ago, stray rocks, boulders, and asteroids were spread all through the inner regions of the solar system. They bombarded the surfaces of the planets. We can still see the results of this bombardment when we look at the Moon and Mercury. Unlike Earth, these worlds have no air or water, so no erosion of their craters has taken place.

MERCURY FACT FILE

Mass: 0.055 Earth, 3.3×10^{23} kg
Diameter: 0.38 Earth, 3,026 miles
Density: 5.43 g/cm^3
Surface temperature: maximum 800°F, minimum –290°F
Rotation period relative to stars (day length): 58.65 Earth days
Distance from Sun (average): 0.387 astronomical units, 36 million miles
Period of orbit (year): 88 Earth days

◁△ Mercury has many craters, formed by collisions with asteroids and meteorites (left). A large impact crater often has a flat floor. Asteroids smash up the surface rocks. Molten lava is then released from inside the planet, and it forms a new surface that is smooth and flat (above).

IMPORTANT DISCOVERIES

1631	Astronomers observe a transit of Mercury across the sun for the first time on November 7. It had been predicted by Johannes Kepler.
1965	The rotation period of 58.65 days is measured by radar.
1974–1975	*Mariner 10* takes the first photographs of the surface.
1985	Sodium is detected in the atmosphere.

VENUS

Venus comes closer to Earth than any other planet. A thick, cloudy atmosphere hides the surface from direct view. Radar images show a great variety of craters, volcanoes, and mountains. The surface temperature is hot enough to melt lead, and the planet may once have had vast oceans.

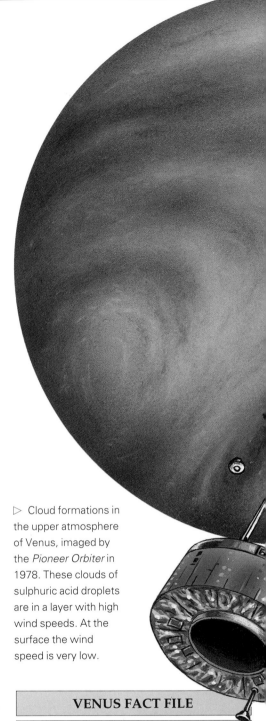

Venus is the second planet from the Sun, orbiting on a nearly circular path almost 67 million miles from the Sun in 225 days. It takes 243 Earth days to rotate, the longest of any planet. Venus actually spins backward, that is, in a direction opposite to its orbital motion. This slow, backward rotation means that, as seen on Venus, the Sun rises and sets only twice a year, because its solar day lasts 117 of our days. Venus comes within 28 million miles of Earth, closer than any other planet ever gets to us.

Venus is only a little smaller than Earth, and has almost the same mass. For these reasons, Venus is sometimes referred to as Earth's twin or sister. However, the two planets have very different surfaces and atmospheres. Earth has rivers, lakes, oceans, and an atmosphere we can breathe. Venus is a scorching hot planet with a thick atmosphere that would be poisonous to humans.

Astronomers knew very little about Venus until the space age. Thick clouds prevent us from observing the surface with telescopes. However, spacecraft have descended through the atmosphere of Venus, which is mainly carbon dioxide, with traces of nitrogen and oxygen. The pale yellow clouds in the atmosphere contain droplets of sulphuric acid, which fall as acid rain.

Observing Venus

Venus is the easiest of all the planets to spot in the sky. Its dense clouds reflect sunlight strongly, making it bright. Because its orbit is closer to the Sun than Earth's, it never gets very far from the Sun as viewed in our sky. For a few weeks every seven months, it is the brightest object in the western sky in the evening. People call it the "evening star." At these times it appears 20 times brighter than Sirius, the brightest star in the northern sky. Three and a half months later, it will be rising three hours earlier than the sun, to appear as a brilliant "morning star" in the eastern sky.

You can observe Venus an hour or so after sunset or an hour before sunrise. The angle between Venus and the Sun is never greater than 47°. The two positions in the orbit where the angle gets this great are known as the greatest eastern elongation and the greatest western elongation. For two or three weeks near the greatest elongations, Venus cannot be missed if the sky is clear. If you can first see Venus in the dawn sky, at the time of greatest western elongation, it is so bright that it should be possible to continue to find it even when the sun has risen. If you use binoculars or a telescope, make sure the sun cannot get into the field of vision by mistake.

It is not difficult to see that Venus has phases, like the Moon. At the greatest elongations, the planet looks like a miniature half-moon. As the planet comes closer to Earth, you will see that the size of the image gets a little larger each day and the shape changes to a narrow crescent. You will not see any surface markings because the clouds are too thick.

▷ Cloud formations in the upper atmosphere of Venus, imaged by the *Pioneer Orbiter* in 1978. These clouds of sulphuric acid droplets are in a layer with high wind speeds. At the surface the wind speed is very low.

VENUS FACT FILE

Mass: 0.815 Earth, 4.87×10^{24} kg
Diameter at the equator: 0.949 Earth, 7,519 miles
Density: 5.25 g/cm^3
Surface temperature: 900°F (maximum)
Rotation period relative to the stars: 243 days
Distance from Sun (average): 0.723 astronomical units, 67 million miles
Period of orbit (year): 224.7 days

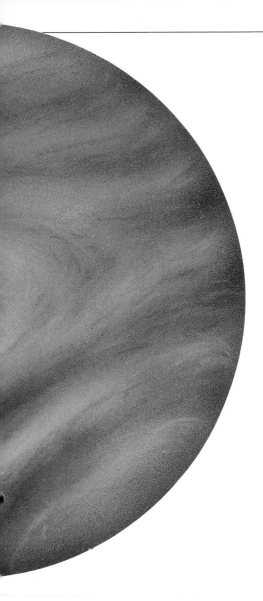

Transits of Venus

On rare occasions, Venus passes directly between Earth and the Sun. These transits, as they are known, were used in the 18th century to determine the size of the solar system. By timing differences between the start and finish of the transit as seen from different places on Earth, astronomers estimated the distance from Earth to Venus. Captain James Cook's third voyage of discovery (1776–79) included transit observations. Venus will next cross the Sun's disk in 2004.

Phases of Venus

The first person to observe the phases of Venus was Galileo in 1610. He realized the moonlike shapes meant that Venus orbits closer to the Sun than does Earth. His observations of Venus proved that the Sun is at the center of our solar system. By observing the phases of Venus every few days for a month or so, you should be able to tell whether it is approaching us or going away.

A torrid world

The atmosphere on Venus is extremely hot and dry. The temperature reaches a maximum of about 900°F at the surface.

NAMING FEATURES

The names of features on planets and moons are decided by the International Astronomical Union. The *Magellan* spacecraft found a large number of craters, mountains, and plains on Venus that required names. Venus was the Roman goddess of love, and scientists decided to name the new features after famous women throughout history.

Several large landmasses are named after goddesses from many different cultures, not just from ancient Roman and Greek myths. For example, Navka Planitia is named for the Arabic mother goddess of good fortune and Sedna Planitia for an Inuit (Eskimo) goddess. Ushas Mons refers to the Indian goddess of dawn and Ozza Mons to a Persian goddess.

Two American women writers, Edith Wharton and the poet Emily Dickinson, have craters named after them on Venus.

◁ Venus orbits closer to the Sun than does Earth. When Venus is on the opposite side of the Sun, the whole disk is illuminated, whereas when it is between Earth and the Sun we see only a part of the sunlit hemisphere. For this reason, Venus shows phases, as the Moon does, as it travels along its orbit.

There is 105 times as much gas on Venus as in Earth's atmosphere. The pressure at the surface is also very high, 95 times greater than at the surface of Earth. Spacecraft have to be specifically designed to survive the crushing atmosphere. In 1970 the first spacecraft survived the tremendous heat for nearly an hour, just long enough to send back a single photograph. Russian landers in 1982 sent back color pictures from the surface, showing sharp rocks.

Venus is very hot because of the greenhouse effect. The atmosphere, a thick blanket of carbon dioxide, traps heat from the Sun. So much heat energy is trapped that the temperature of the atmosphere is much hotter than in an oven.

On Earth, where there is a small amount of carbon dioxide, the natural greenhouse effect raises the global temperature by 85°F. On Venus it produces a temperature 750°F higher. By studying the extreme greenhouse conditions on Venus, we may be able to understand the extra warming effect on Earth caused by the release of carbon dioxide when fossil fuels such as coal and oil are burned.

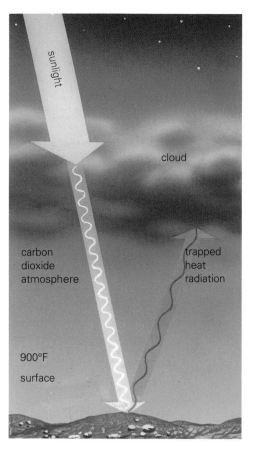

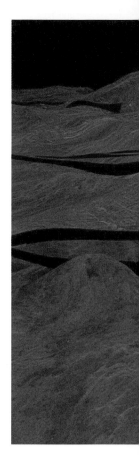

◁ The atmosphere of Venus contains a lot of carbon dioxide. This gas allows heat radiation from the Sun to pass through the atmosphere of Venus to warm the surface of the planet. The heat radiated by the rocks is of longer wavelength than the Sun's rays, and this cannot pass through the carbon dioxide. So, the carbon dioxide acts like the glass in an ordinary greenhouse: the Sun's rays get in, but the heat released is trapped.

Venus and Earth, long ago

When Earth was formed, 4.5 billion years ago, it too had a dense atmosphere of carbon dioxide, just like that on Venus. However, this gas dissolves in water. Earth was cooler than Venus because it is farther from the sun and, as a result, rain washed the carbon dioxide out of our atmosphere and into the oceans. The shells and bones of marine animals created rocks, such as chalk and limestone, in which carbon and oxygen are trapped. The formation of coal and oil also removed carbon dioxide from our atmosphere.

There is not much water in the Venus atmosphere. The greenhouse effect means its temperature is above the boiling point of water up to a height of nearly 30 miles. There may have been oceans on Venus in the past but, if so, they have now boiled away.

The surface

Astronomers have used both visiting spacecraft and radio waves to probe beneath the clouds around Venus. More than 20 American and Russian spacecraft

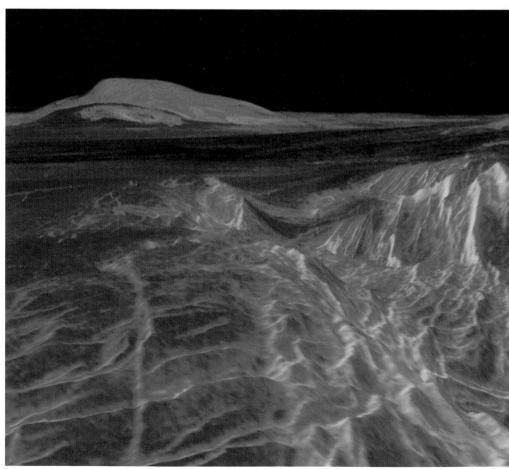

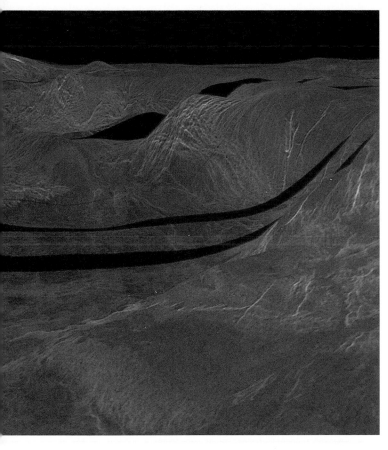

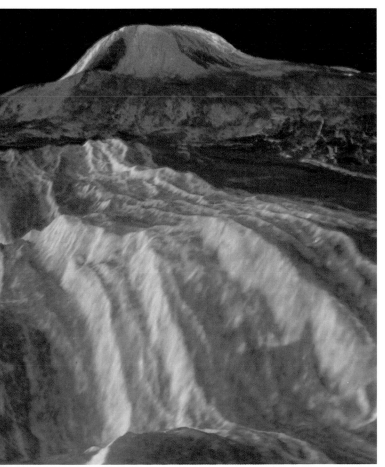

have been to Venus, more than to any other planet. The first Russian craft were crushed by the atmosphere. In the late 1970s and 1980s, however, the first photographs were taken, which showed a variety of sharp and shattered rocks, as well as dust, with a chemical makeup similar to the volcanic rocks on Earth.

In 1961 scientists sent radio waves all the way to Venus and back in order to measure the planet's rotation rate. Radar maps made by radio telescopes on Earth showed that Venus has landmasses the size of continents. About three-quarters of the planet is covered in flat plains, and some large mountains could be picked out.

A spacecraft named *Magellan,* launched in 1989, was sent into orbit around Venus in 1990. By using radar, *Magellan* was able to map details on the surface as small as 300 feet wide. Computers were used to turn the radar information into images that are just like photographs. *Magellan* imaged volcanoes, a remarkable range of mountains, and other features.

Impact craters

Magellan sent back beautiful images of huge impact craters on Venus. These craters formed when giant meteorites plunged through the atmosphere and crashed onto the surface. The impacts also released molten lava trapped inside the planet. Some of the meteorites exploded in the lower atmosphere, creating huge blast waves that left dark, circular craters. A meteorite crashing through the atmosphere travels at about 35,000 miles per hour. When it hits the ground, the rock is turned instantly to vapor (hot gas). The surface rocks are blasted away, scooping out a crater. Sometimes lava that lies many miles below the surface escapes and oozes away from the crater.

Volcanoes and lava

Venus is covered with volcanoes, many hundreds of thousands of them. A few are very large: almost 2 miles high and 300 miles wide. Most of the volcanoes are just 1 to 2 miles across and about 300 feet high. The lava-flows on Venus are longer than any on Earth. Since Venus is too hot for there to be any ice, rain, or storms, there is no weather. This means that the volcanoes and craters have hardly changed in the millions of years since they were formed. The pictures of Venus from *Magellan* show a landscape older than anything you can see on Earth, but it is younger than on many other planets and moons.

Venus appears to have a rocky skin. Below it there are hot currents of lava circulating, which stretch the thin surface. Lava is constantly gushing through holes and rips in the rocky surface. The volcanoes also belch out sulphuric acid droplets. In some locations thick and sticky lava has oozed into puddles up to 15 miles across. There are other places where large bubbles of lava at first formed domes on the surface and later collapsed.

On Earth it is not easy for geologists to determine our planet's history because wind and rain are constantly eroding the mountains and valleys. Planetary scientists are interested in Venus because the surface is like a fossil. The features mapped by *Magellan* are hundreds of millions of years old. The volcanoes and lava-flows are all preserved on this dry planet, which is the closest world to our own.

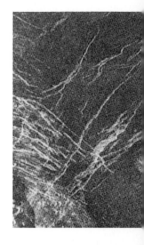

▽ A false-color image taken above the Sapa Mons volcano, which is 250 miles across and 1 mile high. Lava-flows have erupted all around the sides of the volcano. This is typical of large volcanoes on Earth, especially those in Hawaii.

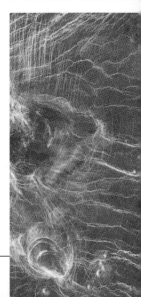

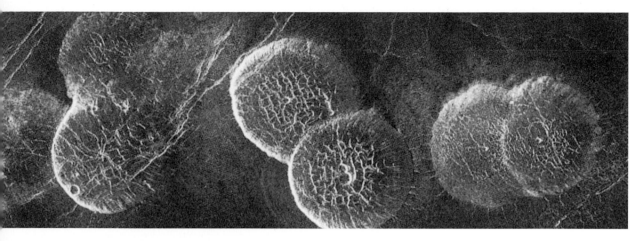

◁ These strange lava pancakes are very thick domes of lava about 15 miles in diameter. Thick and sticky molten rock may have bubbled through the surface cracks in this region.

◁ Three large craters from 15 to 30 miles in diameter on the Lavinia plains of Venus. The dark inner rings are filled with lava. The bright petal shapes around the crater are found in this form only on Venus. They are a jumbled mass of rock that was blasted out of the craters by meteorite impacts.

IMPORTANT DISCOVERIES

1610	Galileo observes and records the phases of Venus.
1639	First observation of a transit of Venus.
1958	Radio measurements suggest a high surface temperature.
1961	Rotation period of 243 days measured by radar.
1962	*Mariner 2* is the first spacecraft to fly by Venus, and confirms the high surface temperature.
1970	*Venera 7* lands and sends data back from the surface.
1974	*Mariner 10* flyby takes 4,000 images of the clouds of Venus.
1978	First detailed maps obtained by *Pioneer* probe, using radar.
1982	Analysis of surface rocks and soil by *Venera 13* and *14* landers.
1990	*Magellan* spacecraft starts radar mapping of the surface in great detail.

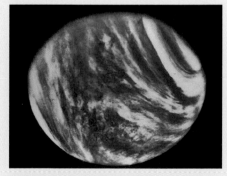

△ The cloudy atmosphere of Venus.

◁ These spidery webs are networks of surface rocks that are cracking. Underneath the circular patterns, hot rock is trying to break through, causing the crust of the planet to bulge. *Magellan* found hundreds of these shattered blobs and domes.

THE MOON

The Moon is only 235,000 miles from Earth. It is the only world in space that people have visited. The Moon has no air, no water, and no weather. Its surface has mountains, craters, seas of solidified lava, and layers of dust. In the next century we may build space stations for scientific work on the Moon.

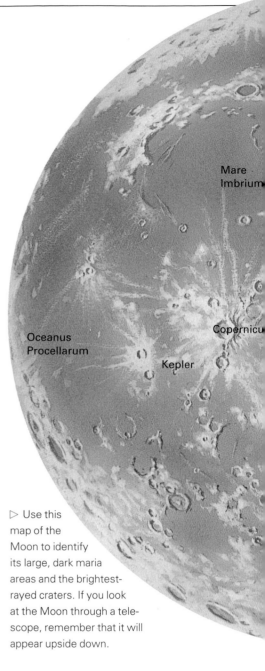

Mare Imbrium

Oceanus Procellarum

Copernicu

Kepler

The Earth is the only planet in the inner solar system with a large moon. Earth has 81 times the mass and nearly four times the radius of the Moon. The pair is held together by the force of gravity. On the lunar surface the pull of gravity is only one-sixth of that on Earth because the Moon has a much smaller mass. Gravity is so weak that any air or water it had has escaped into space.

The Moon keeps the same face to Earth all the time. However, it wobbles a bit, which means nearly three-fifths of the Moon is sometimes visible from Earth. The slight wobble in its motion, which is known as libration, allows us to see slightly more than half the Moon during a month. Photographs taken by orbiting spacecraft have shown that the far side is mainly mountainous.

▷ Use this map of the Moon to identify its large, dark maria areas and the brightest-rayed craters. If you look at the Moon through a tele-scope, remember that it will appear upside down.

◁ This picture of the Moon is made up from two photographs, each taken when only half the Moon was visible from Earth. It shows many details because the Sun was illumi-nating the craters and mountains at an angle and casting long shadows. In a photo-graph taken at Full Moon, there is less shadow to help high-light the lunar features.

Looking at the Moon

You do not need any instruments to start Moon-watching. Try to look at the Moon for about a month and follow the changes in phase. For the few days around the New Moon, you probably will not be able to see the Moon at all. It is easy to see that the Moon always presents the same face to Earth, and that there are bright patches (the mountains) and darker areas, too.

You can see quite a lot using binoculars to look at the Moon. You will easily be able to see dark, flattish areas called *maria*, a Latin name given by Galileo, who likened them to seas on Earth. Maria are huge

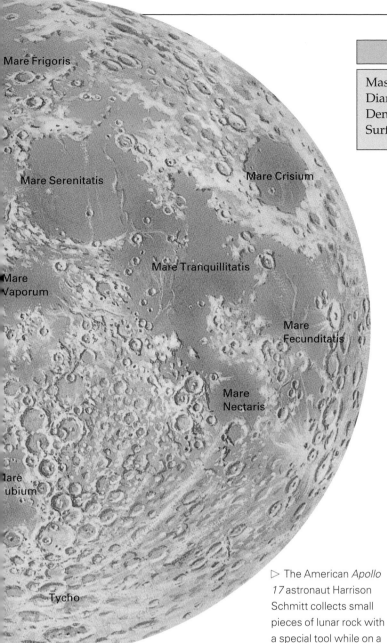

Mare Frigoris

Mare Serenitatis

Mare Crisium

Mare Tranquillitatis

Mare
Vaporum

Mare
Fecunditatis

Mare
Nectaris

Mare
Nubium

Tycho

MOON FACT FILE	
Mass:	0.0123 Earth, 7.35×10^{22} kg
Diameter:	0.273 Earth, 2,160 miles
Density:	3.342 g/cm³
Surface gravity:	0.1653 Earth

▽ Part of the moon's surface photographed from Earth. The Sun is very low in the lunar sky and the sharp contrast between light and shadow emphasizes the craters.

▷ The American *Apollo 17* astronaut Harrison Schmitt collects small pieces of lunar rock with a special tool while on a lunar mission in December 1972.

plains of lava that flowed out of the Moon long ago. You can also see a few large craters, formed when huge asteroids crashed into the Moon billions of years ago. Along the border between the part of the Moon in sunlight and that in darkness, the jagged outlines of mountains can sometimes be glimpsed. What you see is very similar to the views that delighted Galileo when he turned a telescope on the Moon in 1610.

To see more of the lunar landscape, you need to look through a small telescope. The largest craters have rays of bright material stretching like spokes in all directions, and some go nearly one-quarter of

the way around the Moon. A few have a small mountain in the center of the crater. Hundreds of smaller craters can be seen. If you look at the Moon for a week or so, you will see the shadows of mountains and craters change in size.

The surface

The Moon has no atmosphere. Its sky is always black, even in daylight, because there is no air to scatter sunlight and create the blue sky we have on Earth. Sound waves cannot travel in a vacuum, and so the Moon is completely silent. There is no weather: the landscape has not

been shaped by rain, rivers, and ice as it has on Earth. During the lunar day, the surface temperature in direct sunlight goes well above the boiling point of water. Humans who explore the Moon wear air-conditioned spacesuits for protection from the intense heat. At night the Moon's temperature plunges to 300°F below freezing.

Exploring the moon

In 1959 a Russian spacecraft made the first flight past the Moon, and transmitted rather blurred pictures of the far side. Just 10 years later, American scientists achieved the first lunar landing. In all, six Apollo crews visited the Moon between 1969 and 1972. They set up experiments and brought back 850 pounds of lunar rocks and soil.

In the years before the Apollo mission, spacecraft orbited the Moon and photographed it in great detail to locate suitable landing sites. After this, robot spacecraft made the first controlled landings. They tested the strength of the surface, which some scientists believed was a thick dust layer. In fact the landing craft sank only a few inches, showing that people could walk on the Moon without sinking into the dust. The final stages of planning involved sending astronauts into orbit around the Moon in 1968, without attempting a landing.

In the first landing missions the astronauts traveled on foot, staying close to the spacecraft. Later missions included the use of a car with mesh tires, and the astronauts stayed several days on the Moon. The scientific experiments set up on the Moon used several instruments to measure lunar earthquakes (moonquakes). Other experiments studied the matter being blown away from the Sun and the flow of heat from the Moon.

The study of the lunar rocks continued for over 20 years, and a great deal of information was learned about the origin and history of the Moon. All the rocks brought back are made from cooled lava. The dark ones, found on the maria, are like the basalts on Earth. Many of the rocks had been smashed to bits in meteorite explosions, then crushed back together again. None of them has ever been exposed to water, and they contain no fossils. The Moon is a barren place.

The age of the Moon

By looking at radioactive substances in the rocks, scientists have worked out the age of the moon. Uranium, for example, slowly changes to lead. In a sample of uranium 238, half the atoms turn into lead atoms within 4.5 billion years. So, by measuring the proportions of uranium and lead, the age of a rock can be worked out: the more lead there is in the rock, the greater its age.

The rocks on the Moon first became solid about 4.4 billion years ago. The

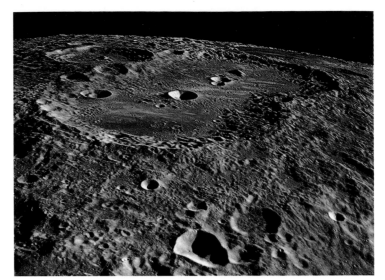

◁ The large crater in this *Apollo 17* photograph of the far side of the Moon was named Van de Graaff in 1970 in honor of an American physicist. It is 151 miles across.

▷ In each of the Apollo lunar missions, one astronaut remained in orbit around the Moon in the Command and Service Module (CSM) while two others landed on the surface in the Lunar Module. This is the *Apollo 16* CSM over the far side of the moon.

Moon's formation must have been a little before this, and an age of about 4.65 billion years is likely. This agrees with the ages of meteorites, and also with the estimated age of the Sun. Before the Apollo mission, all estimates of the Moon's age were just guesses.

Lava-flows and giant impacts

The oldest rocks on the Moon are in the mountain regions. Those from the frozen seas of lava are younger. When the Moon was very young, the outer layer was molten because the rock had been made liquid by heat. As the Moon cooled, a skin, or crust, of rock formed, and parts of this are now found in the lunar highlands. For the next half a billion years, the crust was relentlessly pounded by asteroids, tiny planets, and giant rocks left over from the

formation of the solar system. The largest impacts left huge circular depressions that eventually became the maria. The planets were bombarded as well, but on Earth nearly all trace of this has disappeared because of weathering.

As the outer layers of the Moon continued to cool, the interior gradually warmed up, heated by radioactivity. Between 4.2 and 3.1 billion years ago, lava oozed out through holes in the crust, flooding the circular basins left by giant impacts. The lava was quite liquid and flowed in great sheets to create the lunar maria. So when Galileo named these regions "seas" he was right in a way, for they are solidified oceans of rock that were once liquid. The flow of lava continued for nearly a billion years. We know this simply from looking at the spread in the ages of rocks.

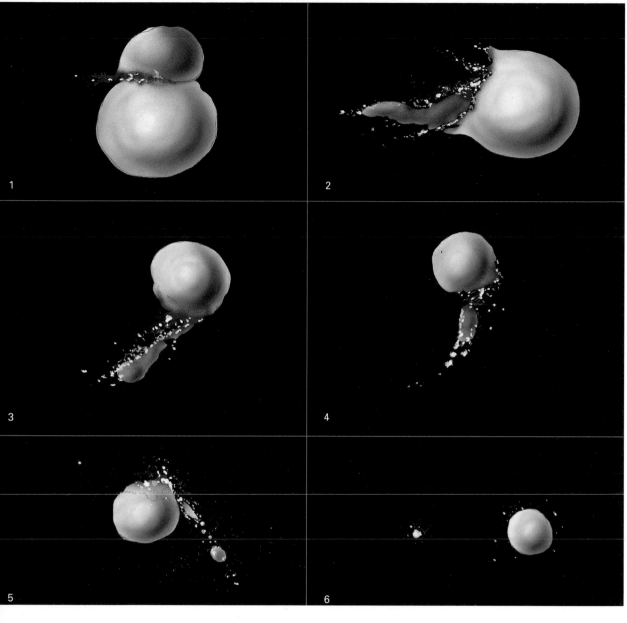

◁ These illustrations explain how Earth might have got its Moon. Scientists can calculate, with the help of a computer, what would have happened if a planet the size of Mars crashed into the side of Earth soon after it formed. A long string of rocky fragments is drawn out from Earth like a tail, while all the iron falls onto Earth and settles in its core. Part of the rocky tail combines to make the Moon.

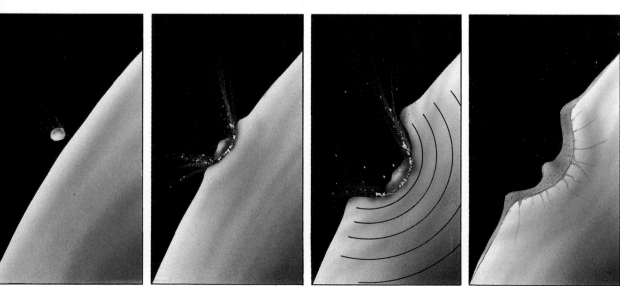

◁ Most lunar craters are created when high-speed rocks crash onto the Moon's surface from space. In the explosion that occurs on impact, the falling rock burns up and a deep circular hollow is gouged in the ground. Rock fragments are blasted out. Some fall back into the crater and some over the surrounding area.

About 2 billion years ago, the volcanic time on the Moon finally came to an end. The outer layers of rock became thick enough to plug the flow of lava. Since then all changes on the surface have been caused by impacts. The larger craters with rays are the result of big impacts, which caused huge explosions at the point of collision; debris was sprayed out radially for hundreds of miles. Endless collisions with smaller rocks have continuously smashed up the surface and scooped out the smaller craters.

Inside the Moon

Apollo astronauts left seismometers (instruments for detecting earthquakes) in four places. These instruments detected gentle moonquakes, nothing as strong as our earthquakes. By observing vibrations from the same moonquake at more than one location, scientists can discover the internal structure of the Moon. The way in which the moonquake waves travel suggest that the lunar crust is between 35 to 60 miles thick. Underneath this there is a 600-mile thick layer of cold, dense rock. Finally, there is a warm core that is partly molten. However, unlike Earth, the liquid core does not have very much iron, so the Moon has no magnetic field.

Where did the Moon come from?

Before scientists looked at lunar rocks, they had three different theories for the Moon's origin but no way of telling which was correct. Some astronomers believed the young Earth had been spinning so fast that it flung off a great glob of matter that became the Moon. Others suggested it had been captured from elsewhere, even though this is highly improbable. A third idea was that Earth and the Moon formed separately, at more or less the same time, and at roughly similar distances from the Sun. Differences in the chemical makeup of the Earth and Moon seemed to indicate that the two objects were never physically united.

Recently, a fourth idea has been accepted as the most likely explanation of the Moon's origin. This is the giant impact hypothesis. The basic idea is that when the planets we know today were still forming, a body the size of Mars slammed at a glancing angle into the early Earth. The lighter material in Earth's outer layers was torn away, forming a ring of debris surrounding Earth, which retained all of its iron core. Eventually, this ring condensed to form the Moon. The giant impact idea explains why Earth has a lot of iron, and the Moon almost none. The impact also removed many of the gases, such as oxygen, from the material that was to turn into the Moon.

Return to the Moon

Astronomers are finding that pollution on Earth is making it increasingly difficult to observe the skies. Lights blaze out from big cities, smoke and volcanic eruptions cloud the skies, and TV stations interfere with radio astronomy. Furthermore, observations cannot be made from the ground of infrared, ultraviolet and X-radiation. A science village on the Moon could

be the next important step in our exploration of the universe.

In many ways the Moon would be ideal for an observatory. Earth-orbiting telescopes, such as the Hubble Space Telescope, are used now to get above the atmosphere, but telescopes on the Moon would be far superior. The far side of the Moon shields instruments from the Earth, and the Moon's slow rotation means the nights last for 14 of our days. This would allow astronomers to make observations of a star or galaxy without interruption for far longer than is possible at present.

Two kinds of astronomy are extremely difficult on Earth: neutrino astronomy and the search for gravity waves. Neutrinos are extremely small particles produced by the Sun and stars. Gravity waves would be created by two black holes orbiting each other, or by explosions in the centers of galaxies. The Moon would provide an isolated environment for conducting difficult observations of all kinds in astronomy. It is for these reasons that astronomers are likely to be the first scientists to return to the Moon.

Lunar resources

If people should ever want to explore space beyond the Moon, then the Moon could be the base station. The low lunar gravity means that it would be 20 times easier and cheaper to launch a huge space station from the Moon than from Earth. Water and breathable gases can be made on the Moon, because there is hydrogen and oxygen in lunar rocks. Abundant supplies of aluminium, iron, and silicon could supply materials for buildings.

A lunar base is essential for finding out more about the materials available on the Moon and for learning how to do engineering and space science in the lunar environment.

Robots for astronomy

The *Clementine* mission put a small satellite in lunar orbit in 1994. This is the first step in restarting research on the Moon after a 20-year delay. The next stage in expanding lunar science is likely to include the construction of automatic and robotic observatories. Astronomers have plenty of experience in operating telescopes in space. Unmanned missions with scientific goals will be the start of lunar exploration and exploitation.

Lunar observatories will be on the Moon's far side, and so satellites will be placed in permanent lunar orbit to beam signals from the lunar far side back to Earth. Lunar telescopes will need to function without human operators, and in extreme temperatures. One idea, being worked on at the University of Arizona, is for a stationary telescope with almost no moving parts. The Moon's own slow rotation and orbital motion would change the direction of view during an 18.6-year cycle. Over time, the telescope could study millions of stars and galaxies.

Very large radio telescopes can be built on the Moon. For the first time it would be possible to observe very long wavelength radio waves from the universe. By combining data from Earth-based and lunar telescopes, it will be possible to see into the central mile or so of the most energetic galaxies in the universe.

◁ Sometime in the future, astronomical observatories may be built on the Moon. This is what a station for observing the Sun from the Moon might look like. In early 1994 the *Clementine* mission mapped the whole of the Moon for the first time, in a series of 3 million photographs. These were used to make maps of the geology of the Moon, from which sites for future exploration and development will be identified.

MILESTONES TO THE MOON	
1609	First drawings by Thomas Hariot.
1837	Detailed maps become available.
1840	J. W. Draper photographs the Moon.
1946	Radar echoes obtained from the Moon.
1959	January 2. First flyby of a lunar probe.
1959	September 13. First crash landing of a lunar probe.
1966	January 31. Russian craft *Luna 9* makes a successful landing.
1968	December. The *Apollo 8* astronauts fly around the Moon.
1969	July 20. Neil Armstrong and Edwin Aldrin land on the Moon and return with the first lunar rocks.
1972	December 11. The final *Apollo* mission.
1994	*Clementine* mission returns 3 million lunar photographs.

MARS

Mars is a planet similar to Earth, although it is smaller and colder. There are deep canyons, giant volcanoes, and large deserts on Mars. Two small moons orbit the Red Planet, as it is known.

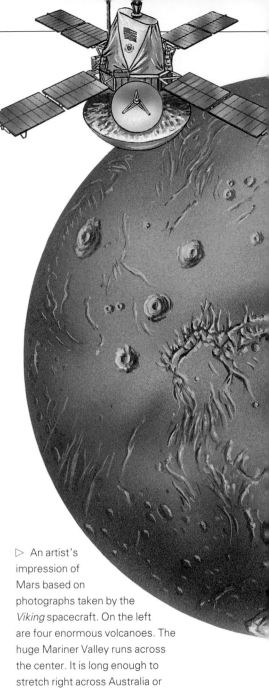

▷ An artist's impression of Mars based on photographs taken by the *Viking* spacecraft. On the left are four enormous volcanoes. The huge Mariner Valley runs across the center. It is long enough to stretch right across Australia or the United States.

Like Earth, the landscape of Mars has been changed by weather, water, and ice. Rivers once flowed on this now dry world. Giant volcanoes, the largest in the solar system, tower above the desert landscape. Huge valleys have opened where the land has cracked. Like the Moon, Mars also has many craters. These are ancient features from a time 3.8 billion years ago when meteorites crashed onto the surfaces of moons and planets.

Mars is the next planet out from Earth, in the direction away from the Sun, and it is the only world beyond the Moon that humans can expect to reach with the rockets that can be made today. For astronauts, this could be the next frontier, to be explored in the 21st century in a journey that will last four years. As the first stage of this international mission, a network of unmanned stations may be set up on the surface of Mars.

If you visited Mars, you would weigh only two-fifths of your Earth weight because Mars has weaker gravity than Earth. You would find the Red Planet (a popular name for Mars) has a thin and wispy atmosphere, made almost entirely of carbon dioxide, with small traces of oxygen and water. Although the atmosphere of Mars is not poisonous, special equipment would be needed to extract enough oxygen for humans to breathe.

Mars takes almost two Earth years to orbit the sun. It has seasons very similar to those on Earth. During the winter periods, astronomers can see ice caps forming in the northern or southern hemispheres. When it is summer, the warm winds stir up giant dust storms across the planet.

Is there life on Mars?

Mars is somewhat like Earth, but much colder. The possibility that there may be life on Mars has long fascinated people. In 1897 the novelist H. G. Wells wrote *The War of the Worlds*, the first book to suggest that Martians existed and that they could invade Earth.

Wells wrote this novel because several astronomers had sketched maps of Mars that showed straight tracks across the planet. In the 1890s at Flagstaff, Arizona, Percival Lowell (1855–1916) made a series of drawings of Mars showing numerous thin straight lines. At the time some observers guessed that these lines were canals, used to transport water from the polar regions to the parched deserts!

The range of temperatures on Mars could be tolerated by some of the simpler plants and creatures that live on Earth. In the summer the surface temperature can rise to just above freezing at midday on the equator. Most of the time, however, the surface temperature is well below zero, but not much worse than in Alaska or Antarctica. Mars has plenty of carbon dioxide in its thin atmosphere, some nitrogen, and traces of water and oxygen. The similarities between the Martian and terrestrial atmospheres led some scientists to believe that Mars could have primitive life.

Spacecraft landers

The science-fiction description of Mars as a planet with advanced life changed abruptly in the 1960s. An American

MARS FACT FILE
Mass: 0.107 Earth, 6.4×10^{23} kg
Diameter: 0.53 Earth, 4,145 miles
Density: 3.95 g/cm^3
Surface temperatures: -9°F overall, -240°F at poles, and 32°F at equator
Rotation period relative to stars (day length): 24.6229 hours
Distance from Sun (average): 1.5237 astronomical units (142 million miles)
Period of orbit (year): 687 days

space probe, *Mariner 4*, produced the first clear pictures of the planet from close up. They showed a lifeless world, shattered by craters.

In 1975, U.S. scientists launched two *Viking* spacecraft. Each carried a lander and an orbiter, which photographed Mars in detail. Almost everything we now know about Mars came from these explorations, which lasted for more than four years. The cameras on the *Viking* landers found no evidence of plants or animals, and the chemistry experiments failed to find the kinds of molecules associated with life. Although Mars is almost certainly a dead world now, we cannot know whether it had primitive life long ago without further exploration.

◁ A drawing of Mars made a century ago by Percival Lowell, working in Flagstaff, Arizona. The long, thin lines are typical of drawings of Mars made in the late 19th century. They were first reported by an Italian astronomer, Giovanni Schiaparelli, who called them *canali*, or natural water channels. This was mistranslated into English as *canals*. By the end of the 19th century, Lowell was reporting 160 Martian "canals," almost all of which were illusions.

▷ The *Viking* lander site. Fine dust lies among the sharp rocks. This red rock is rich in iron. Viking scooped up dust and soil for chemical analysis, to see if any traces of life are present on Mars.

KEPLER AND MARS

For Johannes Kepler the orbit of Mars was the key to unlocking some of the secrets of the solar system. Until Kepler's time, astronomers believed that objects in space moved in circles. This belief made it especially difficult to explain the uneven motion of Mars. Kepler finally succeeded in explaining Mars's motion by using an ellipse instead of a circle. This led him to the three laws of planetary motion, which account for the orbits of all the planets (see page 40).

Giant volcanoes

From the *Viking* photographs, scientists have been able to describe the geology and history of Mars. When the solar system formed, almost 5 billion years ago, Mars was probably like the Moon and Mercury. For the first billion years or so, Mars and other planets were often hit by meteorites, which left numerous craters.

When the planets formed, they were all much hotter than they are today. Mars cooled down more quickly than Earth, because it is smaller. We know that Mars was hot enough in the past to have a liquid interior because there are many volcanoes.

Near the equator of Mars, there are the stupendous volcanoes in the Tharsis region. Tharsis is the name astronomers have given to a bulge on the planet 2,500 miles across and about 6 miles high. This plateau has four volcanoes, each of them gigantic compared to any volcanoes on Earth. They are known as shield volcanoes, and were made from very runny lava that spread far and wide before solidifying. As a result, the Tharsis volcanoes are shaped more like pancakes than cones.

The largest Tharsis volcano, Olympus Mons, towers 17 miles above its surroundings. There are cliffs 4 miles high at its base. Lava and debris have spilled from Olympus Mons to cover an area the size of the state of Arizona. At the top of the mountain, a nest of collapsed craters has left a hole with the same area as the city of Los Angeles.

The highest shield volcanoes on Earth are in Hawaii, where the lava gushing out

◁ The Olympus Mons volcano. There are enormous cliffs at the base of the volcano, which is three times higher than Mount Everest.

◁ The rim of a crater seen from the *Viking* lander site 1. The yellow glow in the sky is light reflected from dust swept up by strong winds.

of the interior has made a series of islands. On Mars a gigantic volcano grew in just one place for as long as molten lava continued to flow.

About two-thirds of Mars consists of highlands with many impact craters, surrounded by debris. In these regions there are branching valleys and landslides. Long ago the surface layer of Mars seems to have been quite wet and the valleys show signs of water erosion.

Deep valleys and canyons

Near the Tharsis volcanoes, a vast system of canyons snakes about one-quarter of the way around the equator. The Mariner Valley is 370 miles wide and so deep that

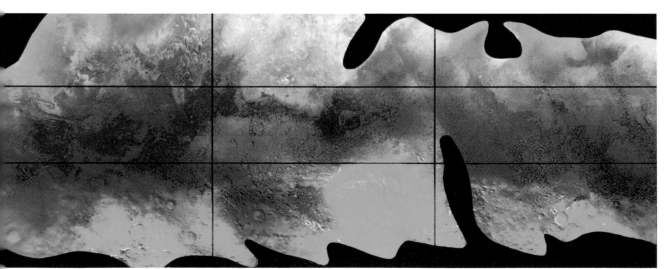

◁ A computer-generated map showing the chemical makeup of the Martian surface. Red depicts rocks with a lot of iron oxide. The dark blue areas are volcanic deposits. Orange and yellow indicate sandy regions and dust. The bright turquoise shows frosty areas.

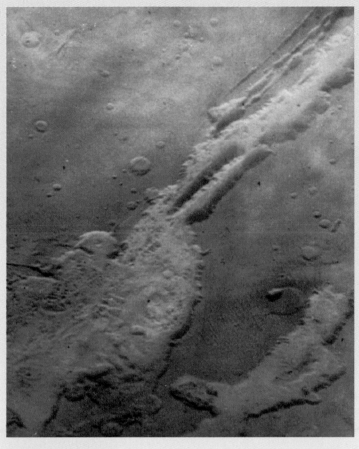

▷ Part of the Mariner Valley, which in this section is 60 miles wide and about 5 miles deep. This huge gash is the result of geological faulting. The steep walls have landslides as well as water drainage channels.

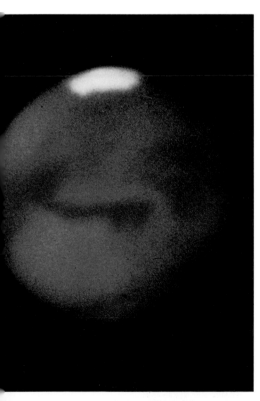

△ The north pole of Mars, seen through a large telescope. The ice is frozen carbon dioxide, sometimes called "dry ice."

designed so they do not topple over if the wheels on one side hit a large stone.

Where has the water gone?

At the north and south poles of Mars there are ice caps. This ice is not the same as the ordinary ice you get in a refrigerator. Mars is so cold in winter, going down to -148°F, that carbon dioxide gas in its atmosphere freezes solid. This kind of ice is known as dry ice. There may be some water ice as well. During the Martian year, the ice caps alternately grow and shrink. Around the poles, there are rings of dunelike features, created by the annual cycle of the ground freezing and thawing.

The thin atmosphere on Mars contains crystals of water ice and very thin clouds. However, even if all the water in the atmosphere fell as rain, it would make a layer only 0.001 inches thick. On Earth the clouds have enough water to make a layer several inches in thickness.

Mars does not have rivers and seas today, but it probably had lots in the past. The *Viking* pictures show plenty of locations where water once flowed. There are islands in wide rivers that are now dry. The chasms of the Mariner Valley, and delicate tracery of dry valleys, were carved by running water that has vanished. Scientists believe that the surface water is now trapped as buried ice, especially in the polar regions. Features like the tundra on Earth, which is frozen soil, could be widespread on Mars.

Mars is a planet where there has been climatic change. The atmosphere is rich in carbon dioxide, which trapped the Sun's heat. The great expulsion of gases from the volcanoes may have added to global warming on Mars as well. All the evidence indicates that in the remote past Mars was a warmer and wetter planet, perhaps with conditions for life to get a foothold. Yet today it is below freezing everywhere.

Dust storms

The surface of Mars is a stony desert. The blocks of stone lying around are volcanic rocks that have been smashed up by earthquakes and eruptions, and by meteorites crashing onto the Martian surface.

Mount Everest would be swallowed by it. Sheer cliffs soar thousands of feet from the valley floor to the plateau above. Deep branching gorges were cut by water that once cascaded across the plateau and into the valley. Chunks of land large enough to hold a major city have slumped into the valley in spectacular landslides. Long ago Mars had huge amounts of water flowing as ground water through the surface layers.

The straight walls of the Mariner Valley suggest that the whole region is quite simply a giant crack, or fault, in the crust of the planet. There are similar valleys on Earth, such as the East African Rift Valley, but ours are much smaller than the Martian ones.

Away from the volcanoes and deep valleys, the landscape merges into the plains in a great jumble of shattered rock. In this region, the enormous lava-flows and giant faults have smashed the crust of Mars and created terrain strewn with boulders that will be a big hazard for the robotic vehicles that will explore Mars in the next century. These will need to be

Between the rocks there is very fine, sandy material. Some of the individual grains of sand are so tiny that you would need a microscope to see them. This sand is more like fine dust than the sand you get on the beach on Earth.

Even though the atmosphere is thin, the winds of Mars easily stir up vast dust storms. The pinkish glow in the Martian sky is real, caused by dust swirling in the atmosphere. Once every two years Mars makes its closest approach to the Sun, and warms up slightly. At that time, known as perihelion, the entire planet can be swamped in dust storms, and no detail can be seen through telescopes on Earth. This is what happened when *Viking* approached Mars in 1976.

The 19th-century astronomers noticed that the color of Mars changes in the course of the year. Some of them thought this could be due to the growth of plants during the Martian summer. We now know that the choking dust storms move sand around. By covering and uncovering darker areas of rock, the storms cause the color changes seen through telescopes.

Space missions

So far there have been 25 spacecraft sent toward Mars. Some never got there and others failed to work. The *Vikings* were a great success in the 1970s, but the failure in 1993 of Mars *Observer* was a big setback.

If Mars had life in the past, there might be very tiny fossils in the rocks. By the end of the century, space scientists hope to use robotic explorers to bring Martian rocks back to Earth. NASA has already developed small Mars robot cars, which would explore the planet under remote control. There are even ambitious plans to land scientists on Mars. This would be an international mission backed by several industrialized nations. However, it would be enormously expensive and quite risky.

Phobos and Deimos

Mars has two tiny moons, named Phobos and Deimos, that were discovered in 1877. These moons are hard to see, even with a large telescope. Phobos is about 14 miles across and Deimos, only 10 miles. These two satellites of Mars have several craters. They may be asteroids that Mars has captured, rather than moons that originally formed close to Mars. Their surfaces are dark, like asteroids, and they have a density similar to that of asteroids. Both moons look like the sorts of rocky lumps that formed in the early solar system, perhaps before the major planets formed.

The small moons in the solar system are not perfectly round. There are two reasons for this. First, they may be fragments from impacts of larger objects smashing together. Second, their own force of gravity is too weak—because they are so small—to squeeze them into a rounder shape.

◁ A view of Mars created by computer-processing 102 separate photographs taken by the *Viking* spacecraft. The center of this view shows the entire Mariner Valley, which is 1,900 miles long and up to 5 miles deep. To the left of the valley are three enormous volcanoes, which appear as dark round spots.

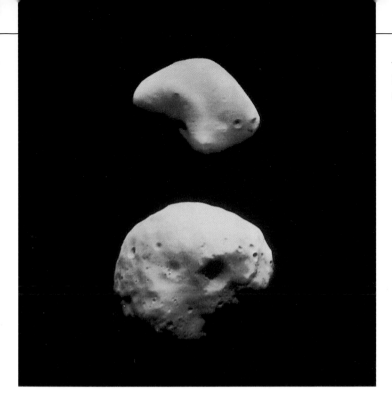

◁ The two moons of Mars, Deimos (top) and Phobos (bottom).

LOOKING AT MARS

Mars is not an easy planet to find in the night sky. You must be patient. Conditions are favorable for about four months once every two years during close encounters called oppositions. When near opposition, Mars is at its closest to Earth. Under the most favorable conditions, which will not occur until August 2003, Mars comes within 34 million miles of Earth. It can be as much as 62 million miles from us at some oppositions, however.

Mars is small, and nothing much can be seen with binoculars, apart from the fact that it is a tiny disk rather than a point of light. To see surface markings, a reflecting telescope of 200 mm aperture or a refracting telescope with a 100 mm lens is the minimum requirement.

At first the small disk of Mars will shimmer around in the heat haze of Earth's atmosphere. Be patient! If you view for 10 minutes or more, you may get a glimpse of the surface if the shimmering stops for a few seconds. With practice, it should be possible to see one of the polar ice caps, which will be whitish, and one or two dusky markings on the surface.

If you can observe Mars with a larger telescope for a few hours, you will see the planet rotate. By observing for a few weeks at opposition, there is a chance that you may see the effects of dust storms. However, this sort of extended observation requires expensive equipment and help from an experienced amateur astronomer.

CLOSE APPROACHES OF MARS

The next few close approaches, or oppositions, will be on these dates:

March 17, 1997   April 24, 1999   June 12, 2001
August 28, 2003   November 7, 2005

IMPORTANT DISCOVERIES

1605 Johannes Kepler proves that the orbit of Mars is an ellipse, with the sun at one focus.	Arizona, mainly in order to observe Mars. Over the next 10 years, he produces maps with extensive canal networks. Lowell says Mars is inhabited and the canals are artificial.
1636 First sketches of surface markings on Mars in an astronomy book.	
1656– Observations show surface markings and a rotation rate of about 24 hours.	**1930** Eugene Antoniadi rejects the suggestion that Mars has canals, following extensive observations.
1666 Cassini measures rotation period of about 24 hours 40 minutes.	**1965** *Mariner 4* returns 21 pictures from Mars showing craters but no trace of canals.
1704 Polar caps discovered.	**1971–** *Mariner 9* returns 7,329 photographs from Mars.
1777– William Herschel suggests the polar caps are thick layers of ice and snow.	**1972**
1837 First detailed maps.	**1976** *Viking 1* lands July 20, followed by *Viking 2* on September 3. The *Viking* orbiters eventually produce detailed maps of most of the Martian surface. The landers investigate the soil in the vicinity of the landing sites without finding traces of life, and send back the first photographs from the surface.
1877 Phobos and Deimos discovered.	
1877 Giovanni Schiaparelli describes a network of about 40 fine lines running across the Martian deserts, and refers to them as *canali*.	
1894 Percival Lowell builds the Lowell Observatory in	

△ Mars, photographed by an amateur astronomer on September 24, 1988, using a 420-mm reflecting telescope. You would be fortunate to see Mars as clearly as this through a telescope. The dark portion extending below the equator is the Syrtis Major, and at opposition it can be seen in a small telescope.

JUPITER

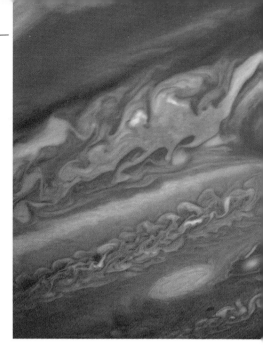

Jupiter is the largest planet in the solar system, 11 times greater in diameter than Earth and 318 times as massive as Earth. It takes about 12 years to orbit the sun, at an average distance of 500 million miles. Belts of clouds in its atmosphere make Jupiter a colorful planet with a Great Red Spot.

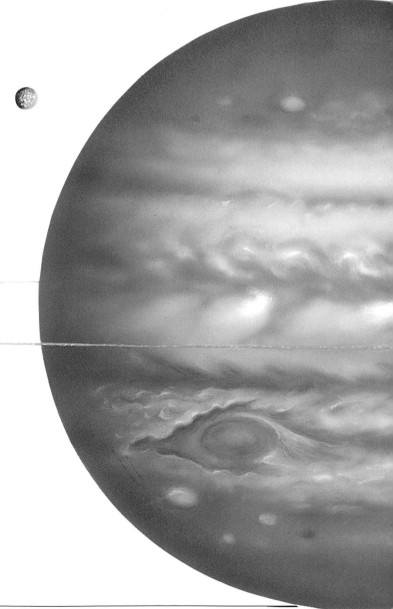

Jupiter is not a solid planet. Unlike the four rocky planets closest to the Sun, Jupiter is a huge ball of gas. There are three more giants farther out: Saturn, Uranus, and Neptune. The chemical makeup of these gas planets is much like that of the Sun and very different from the rocky inner planets. Jupiter's atmosphere, for example, is about 85 percent hydrogen and nearly 14 percent helium. Although we cannot see any solid rocky surface under Jupiter's clouds, deep within the planet the hydrogen is crushed so much that it has some of the properties of a metal.

Jupiter rotates very rapidly, once every 10 hours. This is so fast that the planet bulges at the equator. The rapid rotation also causes high wind speeds in the upper atmosphere, where the clouds are stretched out into colorful bands. Different parts rotate at slightly different rates, and these speed differences cause the bands. The clouds are turbulent, and the appearance of the bands can change in just a few days. There are also a great many whirlwinds and round spots in Jupiter's clouds. The largest of these is the Great Red Spot, which is larger than the earth. It is visible through a small telescope. The Great Red Spot is a huge storm in Jupiter's atmosphere, which has been observed for over 300 years.

There are at least 16 moons in orbit around Jupiter. One of them, Ganymede, is the largest satellite in our solar system, bigger than the planet Mercury.

▷ Jupiter, the giant planet, shown in its true colors in an artist's impression of images returned by the *Voyager* spacecraft. The colored bands are clouds with crystals of frozen ammonia, and chemical compounds of carbon, sulphur, and phosphorus. The Great Red Spot is the oval cloud below the equator and to the left.

◁ The Great Red Spot on Jupiter is a hurricane that must have been blowing for at least 350 years. It is larger than Earth. The white oval below the Red Spot is another whirlwind.

Voyaging to Jupiter

Five spacecraft have already been sent to Jupiter. The fifth, *Galileo*, was launched on a six-year journey in October 1989. The *Pioneer 10* and *11* spacecraft were the first to make measurements. They were followed by two *Voyager* spacecraft, which took spectacular closeup photographs in 1979. After 1991, the Hubble Space Telescope started photographing Jupiter. These images are of similar quality to *Voyager*'s. The Hubble Space Telescope, however, will take photographs for many years, whereas the two *Voyagers* were each limited to a quick look during a brief flyby.

JUPITER FACT FILE

Mass: 318 Earth, 1.9×10^{27} kg
Diameter at the equator: 11.2 Earth, 89,330 miles
Density: 1.31 g/cm³
Temperature of cloud top: -255°F
Rotation period: 9.93 hours
Distance from Sun (average): 5.203 astronomical units, 482 million miles
Period of orbit (year): 11.86 years

OBSERVING JUPITER

It takes Jupiter 13 months to move around the sky and, for several months each year, it is easily visible for some hours before and after midnight. It shines with a steady whitish light. You can use a monthly sky guide to find the best times to see it and to find its position. You only need to know which direction to look in because Jupiter is one of the brightest objects in the sky.

If you have the chance, look at Jupiter through binoculars. If these can be mounted on a photographic tripod, you will find it easier to observe. The view through even the simplest binoculars is amazing. Seen through a telescope, Jupiter is a wonderful sight: there are dark bands encircling the globe and sometimes darker spots. The planet's cloud belts and the Great Red Spot can be seen through a telescope, but not through binoculars. It takes practice to observe the belts and to identify the Great Red Spot. If Jupiter is observed when the Red Spot is visible, it is possible to tell within an hour or so that the planet is rotating very fast, as the Spot will move across the disk. The Great Red Spot varies in the intensity of its color and the ease with which it can be seen.

△ This photograph shows the belts and clouds of Jupiter that can be seen through a small telescope. With binoculars, only the darkest band would be visible.

Clouds of poisonous gas

The dark, reddish bands on Jupiter are called belts, and the light-colored bands are zones. The spacecraft and Hubble Space Telescope photographs show that there are noticeable changes in the belts and zones over a period of only a few weeks. This is because the features we can see are in fact colored and white clouds in the upper atmosphere. In the vicinity of the Great Red Spot the clouds form beautiful whirlpools and wave patterns. The swirling clouds are blown along by winds of over 300 miles per hour.

Most of Jupiter's atmosphere would be deadly to humans. In addition to abundant hydrogen and helium, there is methane, poisonous ammonia, water vapor, and acetylene. You would find it a smelly place. These gases are similar to those from which the Sun formed.

The white clouds contain crystals of frozen ammonia and frozen water. The brown, red, and blue clouds may be colored by chemicals similar to dyes, or by sulphur. Lightning storms flash through the outer atmosphere.

The active cloud layer is quite thin, less than one-hundredth of the planet's radius. Beneath the clouds, the temperature rises steadily. Although it is -255°F at the cloud tops, only 35 miles down through the atmosphere, the temperature is as high as on the surface of the Earth. Just a little farther in, the temperature of the boiling point of water is reached.

△ Jupiter is regularly photographed by the Hubble Space Telescope, which is able to keep track of the changing pattern of bands and zones.

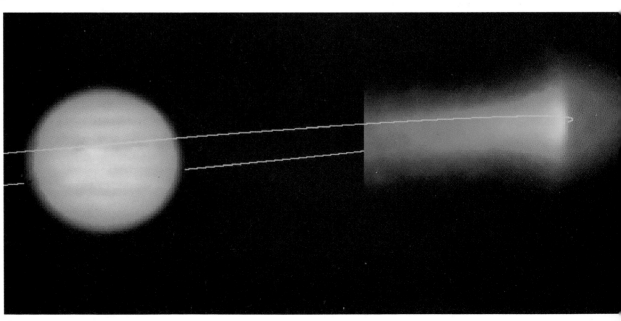

▷ As Jupiter's moon Io orbits, it spews out a cloud of sulphur that encircles the planet. Io's orbit is shown by the yellow line in this composite photograph. To the right, a separate photograph shows the eerie green and purple glow of the sulphur gas.

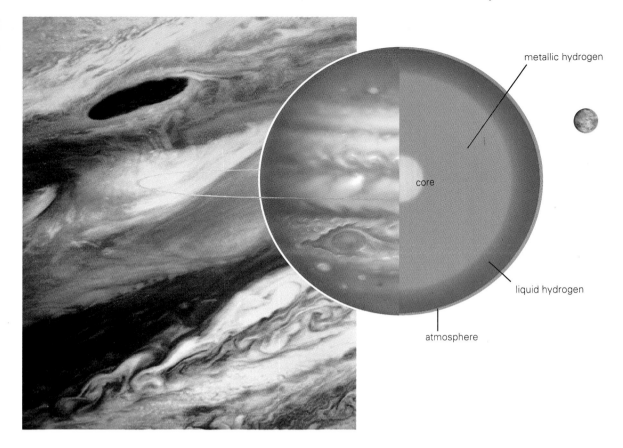

metallic hydrogen

core

liquid hydrogen

atmosphere

◁ Telescopes and space probes can only image the poisonous cloudy layers of the upper atmosphere of Jupiter. Below the thick clouds, hydrogen is squeezed so much that it starts to behave like a liquid metal, and it conducts the natural electricity flowing in the giant planet. At the center of the planet there may be a dense rocky core 20 times as massive as Earth.

Strange matter

Deep inside Jupiter, matter begins to behave strangely. Although there may be a small core of iron at the center of Jupiter, most of the interior is hydrogen. This turns from gas to liquid under the tremendous pressures inside the planet. At deeper and deeper levels, the pressure rises steadily because of the great weight of atmosphere above.

At a depth of about 600 miles, there is a vast ocean of liquid hydrogen. Below 10,000 miles, this hydrogen is being crushed so hard that its atoms are destroyed. The hydrogen then behaves like a metal because it conducts electricity easily. Electricity flowing in the metallic hydrogen produces a strong magnetic field around Jupiter.

The metallic hydrogen in Jupiter is an example of a strange type of matter that can be studied by astronomers but that is nearly impossible to make in a laboratory.

Nearly a star

Jupiter emits more energy than it receives from the Sun. Spacecraft measurements have shown that Jupiter radiates about 60 percent more in heat energy than it gets from solar radiation.

The extra heat is believed to come from three sources: heat still remaining from the formation of Jupiter, energy released by the slow contraction of the planet, and energy from radioactive decay. However, the heat does not come from turning hydrogen into helium, which is what stars do. In fact, the smallest hydrogen-burning stars are about 80 times more massive than Jupiter. This means that other solar systems may have planets that are much larger than Jupiter, but smaller than stars.

Radio station Jupiter

Jupiter is a natural radio station. You cannot make any sense of Jupiter's radio signals because they consist entirely of noise. Electrons whizzing through Jupiter's huge magnetic field make the radio signals. Immense thunderstorms and flashes of lightning add to the radio din. Jupiter has a strong magnetic field, which extends to 50 Jupiter diameters from the planet. No other planet in our solar system has such a strong magnetism and radio emission.

SEE FOR YOURSELF

With a small telescope, or powerful binoculars, you can repeat the observations of Jupiter's four large moons first made by Galileo in January 1610. Close to the planet you will see up to four points of light, like stars. If you observe every night, you will see that they change positions. You are looking at the four largest moons of Jupiter. They are the Galilean satellites, and their names in order of distance from Jupiter are Io, Europa, Ganymede, and Callisto.

Galileo tracked the positions of the four moons for two months. Observing them orbiting Jupiter, he decided they were like a solar system in miniature. This provided important evidence in favor of the theory of Copernicus (see page 23), and it changed astronomical thinking about the solar system.

See if you can observe Jupiter's satellites for a few weeks. You will need some luck with fine weather, but try to plot the positions of the satellites that are visible on as many evenings as possible. Then you can estimate the orbital periods of each moon. For example, Io takes 42 hours to orbit Jupiter, whereas Callisto takes just over two weeks. You will find that the farther a satellite is from Jupiter, the longer it takes to orbit.

It is also possible to observe eclipses of Jupiter's moons. Jupiter casts a shadow into space, and as the moons move through this shadow they disappear from our view. In the 17th century the first accurate measurement of the speed of light was made by timing the eclipses of the moons of Jupiter.

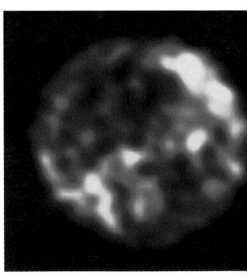

△ Io imaged by the Hubble Space Telescope from a distance of 370 million miles.

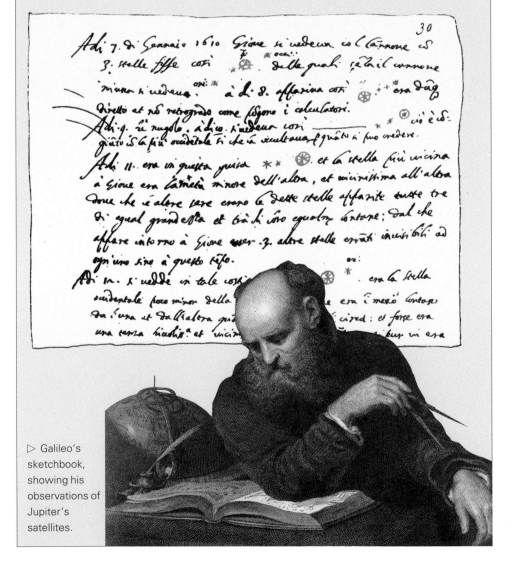

▷ Galileo's sketchbook, showing his observations of Jupiter's satellites.

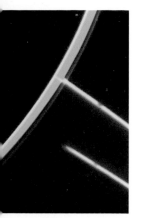

◁ Jupiter has a narrow ring of dust in orbit, discovered by the *Voyager* probes of the 1970s.

Jupiter's moons

Jupiter's family of 16 moons is a solar system in miniature, with Jupiter as the sun and its moons as the planets. The largest moon is Ganymede, 3,270 miles in diameter. It is covered in a thick crust of ice that lies over a rocky core. There is plenty of evidence of bombardment by meteorites, as well as a giant impact from an asteroid 4 billion years ago.

Callisto is nearly as large as Ganymede, and it is heavily cratered. It is the darkest satellite belonging to Jupiter.

Europa has the brightest surface. It is one-fifth water, which has formed a coating of ice 60 miles thick. This reflects light as well as the clouds of Venus.

The most spectacular moon is Io, orbiting close to Jupiter. Io is an extraordinary color, a mixture of black, red, and yellow. Its unusual colors are caused by sulphur that has spewed out of Io's interior. The *Voyager* cameras saw several volcanoes in action on Io, sending fountains of sulphur 100 miles above the surface. At ground level the sulphur lava shoots out as fast as 3,000 feet per second. Some of this matter escapes from Io and now lies in a ring that encircles Jupiter.

The surface of Io is young. We can tell this because there are almost no meteorite craters. Io orbits less than 250,000 miles above Jupiter. It is wrenched by enormous tidal forces. The constant pulling and pushing of tides inside Io cause intense internal friction. This is keeping its interior hot and molten, even though it is far away from the Sun.

Smaller worlds orbit Jupiter

In addition to the four large moons, Jupiter has tiny "moonlets." There are four that orbit even closer than Io, and these are believed to be large lumps of other satellites that no longer exist. The miniature moons orbiting far from Jupiter are probably asteroids that strayed too near to Jupiter and got trapped in its gravitational field. Not all of them have been photographed in detail.

Jupiter also has three very faint rings, first discovered by *Voyager 1*. They are made up of very tiny particles of dust.

The *Galileo* craft will investigate Jupiter and its moons after late 1995. On the larger moons, planetary scientists would like to identify landing sites that could be used by automatic and robotic spacecraft.

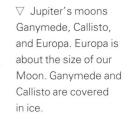

◁ Io is an extraordinary moon. The surface is covered with sulphur that has been thrown out of its active volcanoes.

▽ Jupiter's moons Ganymede, Callisto, and Europa. Europa is about the size of our Moon. Ganymede and Callisto are covered in ice.

Ganymede

Callisto

Europa

SATURN

Saturn, the sixth planet from the Sun, has a wonderful ring system. The density of this gas giant is less than that of water, so in an enormous ocean it would float.

Saturn orbits the sun nearly 10 times farther away than does Earth. This means it receives only one-hundredth as much heat and light as Earth. Consequently, it is a frigid world, with cloud and wind systems similar to those on Jupiter. Saturn takes 29.5 years to orbit the Sun. It spins on its own axis in just 10 hours. This fast spin makes the planet bulge at its equator.

Saturn is 95 times more massive than Earth, and is the second largest planet after Jupiter. Like Jupiter, it is almost entirely made up of hydrogen and helium and pale zones of ammonia clouds. Winds near the equator blow at 1,100 miles per hour, four times faster than Jupiter's worst winds, and 20 times as bad as a severe gale on Earth. The markings are much fainter than those on Jupiter. Occasionally, white spots can be seen by the Hubble telescope.

Saturn's rings

Saturn's beautiful rings do not touch the planet. Three main rings are visible through large telescopes. However, the *Voyager* photographs show that Saturn really has huge numbers of ringlets that blend together when viewed from as far away as Earth. The rings are tilted at an angle of 29° to the orbit. This means that they slowly change their appearance when viewed from Earth. For a year or so we see the rings as wide as they can be in a telescope; then they slowly seem to fold up until 15 years later we view them from the side, and they almost vanish.

The rings are not solid. In fact, bright stars shine right through the rings without their light really being dimmed at all. Although the rings are 250,000 miles across, they are only a few tens of feet thick! The inner parts orbit Saturn faster than do the outer parts.

The rings are mainly made of billions of tiny particles, and each speck is orbiting Saturn like a microscopic moonlet! The particles are probably made of water ice or bits of rock covered in ice. Mostly, they are about three feet in size, but the range runs from a few inches to tens of miles. There are a few larger objects in the rings, stones and boulders up to 500 feet across. They all orbit the planet as if they were unrelated satellites.

The *Voyager* spacecraft focused on Saturn's rings in 1980, and found that there are countless very thin ringlets. The gaps between the rings are caused by Saturn's many moons, which cause the rings to split through the action of gravity.

Why do some planets have rings?

All the gas giants—Jupiter, Saturn, Uranus, and Neptune—have rings. For each of these planets, the rings are found close to the planet. No large moons orbit inside the rings.

A large rocky moon is pulled by enormous tidal forces if it wanders too close to the parent planet. The side of the moon nearest the planet feels a stronger pull of gravity than the far side. The nearer the moon is to the parent planet, the greater the difference in the near-side and far-side gravitational force. An orbit exists close to the planet where these differences, or tidal forces, are strong enough to shatter

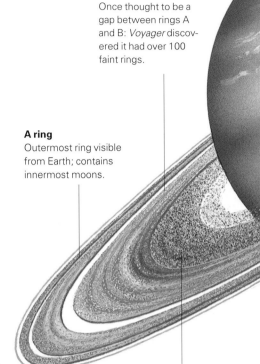

Cassini division
Once thought to be a gap between rings A and B: *Voyager* discovered it had over 100 faint rings.

A ring
Outermost ring visible from Earth; contains innermost moons.

C ring
A blue ring; it is the faintest ring visible from Earth.

▽ ▷ Saturn's rings (right). Planetary rings are thought to form if a moon strays too close to its parent planet and gets shredded by tidal forces (below).

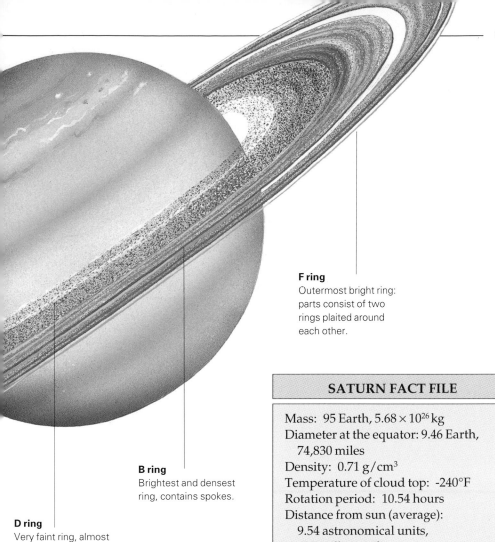

F ring
Outermost bright ring:
parts consist of two
rings plaited around
each other.

B ring
Brightest and densest
ring, contains spokes.

D ring
Very faint ring, almost
touches Saturn.

ordinary rock. So the rings may result when a moon gets shredded up by the gravity of its parent planet. It is also possible that the rings are material left over from the original formation of the planet and its moons.

Titan

Saturn's largest satellite, Titan, is bigger than the planet Mercury. Astronomers think this moon is made of equal amounts of rock and water ice. What is really remarkable, however, is that it has a thick atmosphere, which mainly consists of nitrogen, with some methane (which on Earth occurs as natural gas). No other moon in our solar system has an atmosphere. The pressure at the surface is not much greater than on Earth, but the temperature is only -290°F. However, at this temperature methane exists as a gas, liquid, or solid, depending on local conditions. So Titan could be somewhat like Earth: with rain and snow and oceans and rivers. The difference is that on Titan, these are all made from methane rather than water.

SATURN FACT FILE
Mass: 95 Earth, 5.68×10^{26} kg
Diameter at the equator: 9.46 Earth, 74,830 miles
Density: 0.71 g/cm^3
Temperature of cloud top: -240°F
Rotation period: 10.54 hours
Distance from sun (average): 9.54 astronomical units, 483 million miles
Period of orbit (year): 29.46 years

◁ Titan has a thick, smoggy atmosphere, with several layers of haze.

◁ Dione is one of Saturn's medium-sized satellites. It is 700 miles in diameter and is heavily cratered.

IMPORTANT DISCOVERIES

1610	First telescopic observation by Galileo. His telescope was not good enough to show the rings, and he wrote that Saturn is composed of three parts.
1633	Earliest drawing of Saturn.
1655	Christiaan Huygens discovers Titan.
1656	Huygens announces that Saturn has a ring.
1675	Discovery of the Cassini division in the rings.
1837	Discovery of Encke's division.
1876	Prominent white spot discovered.
1932	Ammonia and methane detected in atmosphere.
1979	*Pioneer 11* encounter.
1980	*Voyager 1* obtains images of Saturn and Titan.
1981	*Voyager 2* mission.
1990	Observations with the Hubble Space Telescope.

Uranus and Neptune

Uranus and Neptune are giant planets made of gas, with very thin ring systems. Uranus is tipped over on its side. Neptune has a stormy atmosphere. Its moon Triton has volcanoes that spew out water and ice.

On March 13, 1781, William Herschel (1738–1822) accidentally discovered a new planet while using a homemade telescope. Herschel was a musician, who settled in Bath, England, where he was an organist. His hobby was astronomy. He made his own telescopes and compiled lists of pairs of stars that appeared very close to each other when viewed through the telescope. One night he saw a new object, which he thought was a comet because it moved slowly against the background stars. Within a few weeks, however, it became clear that this was not a comet but a new planet in our solar system.

Herschel's discovery made him world-famous and King George III paid him a royal pension. At first astronomers could not agree on a name for the planet, but they eventually chose Uranus. In classical mythology, Uranus is the grandfather of Jupiter.

Another new planet, Neptune, was found in 1846 as a result of a careful search. For many years astronomers had been puzzled because Uranus kept drifting off course. They used Newton's law of gravity to work out where Uranus should be, but kept finding that its actual position did not match the predicted one. They knew that this would happen if the gravity of an undiscovered planet was tugging at Uranus.

Two mathematicians set to work on computing where the unseen planet would have to be. In 1845, in Cambridge, England, John Couch Adams (1819–92) teamed up with James Challis (1803–62) at the university observatory. Although Challis actually recorded the new planet,

he did not realize he had found it! At almost the same time, a French astronomer, Urbain Leverrier (1811–77), was trying to persuade the Paris Observatory to search. He also wrote to an observatory in Berlin, Germany. On the night of receiving that letter (September 23, 1846), Johann Galle located the suspected planet. It was named Neptune after the Roman god of the sea.

Uranus—the toppled planet

Uranus is mainly made of hydrogen and helium, but one-seventh of its atmosphere is methane. This gas makes it appear bluish through a large telescope, a fact first noted by Herschel. The space probe *Voyager 2* detected just a few streaks of cloud in the upper atmosphere. The planet's temperature is nearly -365°F. Its center has a large core of rock and iron.

The spin axis of Uranus is tilted over by more than a right angle, which means its north pole actually points below the plane of the planet's orbit. This is unique in our solar system. Uranus takes 84 years to orbit the Sun. Its seasons must be very strange. For about 20 years the north pole points more or less toward the Sun, while the south pole is in permanent darkness.

Astronomers suspect that Uranus may have had a collision with another big planet soon after the formation of the solar system. This could have knocked it onto its side.

Rings around Uranus

The rings around Uranus were found by sheer chance. Astronomers wanted to

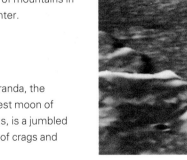

▷ Neptune's largest moon is Triton. This *Voyager 2* image shows a large crater on Triton with a small range of mountains in its center.

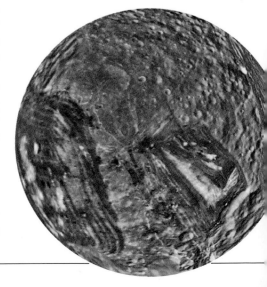

▽ Miranda, the smallest moon of Uranus, is a jumbled mass of crags and cliffs.

URANUS AND NEPTUNE FACT FILE

	Uranus	Neptune
Mass:	14.5 Earth, 8.7×10^{25} kg	17.2 Earth, 1.0×10^{26} kg
Diameter at the equator:	4.0 Earth, 31,900 miles	3.9 Earth, 30,700 miles
Density:	1.27 g/cm^3	1.77 g/cm^3
Temperature:	$-365°F$	$-350°F$
Rotation period :	17 hours 14 minutes	17 hours 52 minutes
Distance from Sun : (average)	19.2 astronomical units 1.78 billion miles	30 astronomical units 2.8 billion miles
Period of orbit (year):	84 years	165 years

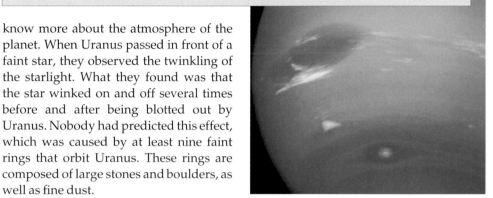

△ A large dark spot on Neptune, as seen by *Voyager 2*.

◁ Uranus (left) and Neptune (right) are four times larger in diameter than Earth. The bluish-green color of their atmosphere comes from the gas methane.

know more about the atmosphere of the planet. When Uranus passed in front of a faint star, they observed the twinkling of the starlight. What they found was that the star winked on and off several times before and after being blotted out by Uranus. Nobody had predicted this effect, which was caused by at least nine faint rings that orbit Uranus. These rings are composed of large stones and boulders, as well as fine dust.

Miranda

There are five large moons and 10 tiny moonlets around Uranus. The most amazing of these is Miranda, which is nearly 300 miles across. The surface has a bewildering variety of valleys, grooves, and steep cliffs. The moon seems to be made of three or four huge lumps of rock that have fused together. It may be the remains of a moon that once crashed with an asteroid and has now managed to pull itself back together again through the force of gravity.

Neptune from *Voyager 2*

Voyager 2 swept past Neptune on August 24, 1989, after a journey of 12 years. Its findings provided many surprises. Because Neptune is 30 times farther from the Sun than is Earth, the sunlight that reaches it is feeble and Neptune's temperature is $-350°F$. However, this is a little warmer than Uranus, which is closer to the Sun. The explanation is that Neptune has internal heat, and it actually gives off nearly three times as much heat as it gains from the Sun.

There is weather in Neptune's atmosphere. *Voyager 2* observed a Great Dark Spot, which seems to be like Jupiter's

Great Red Spot. There are also wispy cirrus clouds. Some of the clouds are frozen methane.

Voyager 2 is now travelling to the edge of the solar system. It will not go to Pluto, the last planet, but astronomers can keep in radio contact until at least 2020. During this time *Voyager 2* will send information on the gas and dust in the far solar system.

Triton

There is a satellite of Neptune larger than Earth's moon: this is Triton. It has an atmosphere of nitrogen, just like Earth's, and is made of seven-tenths rock and three-tenths water. Near Triton's south pole, *Voyager 2* took pictures of red ice, and on the equator it photographed blue ice of frozen methane.

There are huge cliffs carved from water ice, and numerous craters. Neptune deflects comets off course as they come into the solar system. Some of these may have hit Triton, and these impacts could have caused the craters. There are dark streaks of volcanic material. Scientists believe that ice, made of water, methane, and nitrogen, escapes from the interior of Triton through volcanoes.

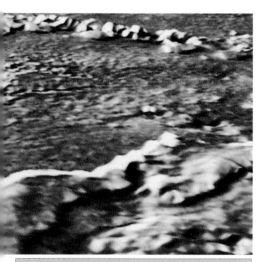

IMPORTANT DISCOVERIES

1690	Uranus first recorded, but as a star.
1781	Uranus discovered as a planet by William Herschel.
1787	Herschel finds two of the satellites of Uranus.
1846	Neptune discovered.
1977	Rings of Uranus discovered.
1986	*Voyager 2* encounter. Ten new moons of Uranus discovered.
1989	*Voyager 2* encounter with Neptune, discovers rings.

PLUTO AND CHARON

At the edge of our solar system, Pluto and its moon Charon together make a double planet. Pluto is smaller than our Moon and is made of rock and ice. There are also numerous objects left over from the formation of the solar system.

Neptune was discovered because astronomers checked carefully on small deviations· in the orbit of Uranus. Early in the 20th century, astronomers observed the orbit of Neptune and came to the conclusion that there might be an even more distant planet. For more than 20 years they searched unsuccessfully. Then in 1930, Clyde Tombaugh, a young astronomer at the Lowell Observatory in Arizona, announced the discovery of a very faint planet, the result of a very detailed search. The diameter of Pluto is 1,400 miles. It has an elongated orbit. The closest Pluto gets to the sun is 2,750 million miles, and the farthest it gets from the sun is 4,583 million miles. In 1979 Pluto actually crossed over Neptune's orbit, and it will pass beyond Neptune again in 1999.

Photographs taken in 1978 showed images of Pluto shaped like an egg. This seemed to indicate that there could be a moon orbiting close to Pluto that the telescopes could not quite image clearly. The sharpest views clearly showing the moon Charon orbiting Pluto have been taken with the Hubble Space Telescope. Pluto and Charon are separated by less than 12,000 miles, and they are like a pair of dwarf planets. Astronomers calculated the mass of Pluto at just 0.0022 (or 1/440) of Earth's mass.

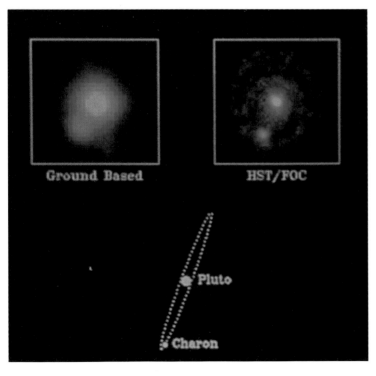

Ground Based

HST/FOC

Pluto

Charon

◁ This photograph, taken with the Hubble Space Telescope, shows Pluto and its moon Charon (at the lower left). The diagram shows Charon's orbit around Pluto as seen from Earth when the photograph was taken.

▷ Pluto and Charon are too far away for any details to be visible, even through the most powerful telescopes, and Pluto is the only planet in the solar system never visited by a spacecraft. Based on what is known about them, an artist has imagined how they might appear from close by when Charon is about to cross in front of Pluto.

The frozen wastes

Pluto is extremely cold. In winter its surface temperature is only –380°F. At closest approach to the Sun, when it is still 30 times farther away than Earth is, the temperature reaches about –330°F. From Pluto our Sun would look like a rather bright star, and you would just be able to see it as a disk rather than a point of light.

Pluto has a very thin atmosphere that probably freezes onto the ground in winter. A thick layer of water and methane ice coats the rocky core. Pluto is unlike planets such as Earth, which are denser and contain iron and nickel. It is completely different from its neighbors, the gas giants such as Neptune. So what is it?

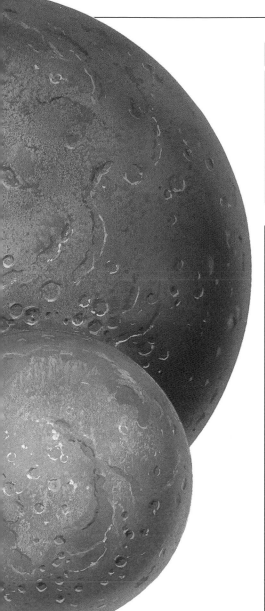

PLUTO AND CHARON FACT FILE

	Pluto	*Charon*
Mass:	0.0022 Earth, 1.3×10^{22} kg	0.0003 Earth, 1.8×10^{21} kg
Diameter:	1,400 miles	750 miles
Density:	2 g/cm^3	2 g/cm^3
Temperature:	−380°F	−380°F
Distance from Sun:	between 29.65 (nearest) and 49.28 (farthest) astronomical units, on a very elliptical orbit	

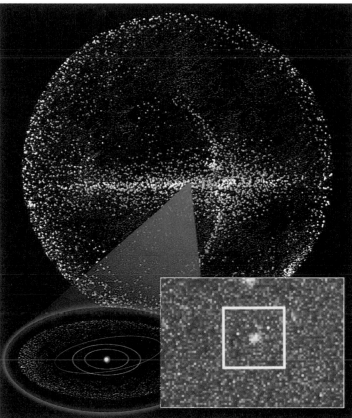

◁ Beyond the orbit of Neptune there is a ring of small planetlike objects of rock and ice, known as the Kuiper Belt (below left). Much farther out, in a huge ball-shaped swarm called the Oort Cloud, there are thought to be billions more similar objects. When these objects are diverted onto paths in our part of the solar system, they become comets.

◁ This object, discovered in 1993, is one of the few known objects in the Kuiper Belt. Called 1993 RO, it orbits the Sun at a distance slightly greater than Neptune's. It is believed to be about 90 miles across.

The Kuiper Belt

At the far edge of the solar system, beyond Neptune, is a ring of small planetary objects known collectively as the Kuiper Belt. Several objects that are from 60 to 120 miles in size have been found, and Pluto could be orbiting within this belt. Charon could have been captured by Pluto from the belt, and perhaps Triton was captured by Neptune. The planetesimals, as the mini-planets are called, are the remnants of the materials from which the larger planets formed in the early solar system.

The chemical makeup of Pluto and Charon is similar to that of comets, but Pluto and Charon are much larger than comets. The smaller planetesimals also seem to have surface materials that are similar to those of the comets.

The search for Planet X

It seems that all large objects inside the orbit of Pluto have now been discovered. The objects in the Kuiper Belt are only the size of asteroids. Much farther out, about a light-year from the sun, we would come to the Oort Cloud of comets. We can be sure there are countless small icy objects to be found, but no major planet. Although careful searches have been made from time to time, astronomers now consider that there cannot be any true planet beyond Pluto. The search for the elusive tenth planet—Planet X—has been in vain.

IMPORTANT DISCOVERIES

1877 First search, made at U.S. Naval Observatory in Washington, D.C.
1905 Careful photographic searches made by Lowell in Flagstaff, Arizona.
1919 Search using what was then the world's largest telescope at Mount Wilson, California.
1930 Pluto detected by Clyde Tombaugh in January. The announcement was made on March 13, 1930, exactly 149 years after the discovery of Uranus.
1978 Pluto's moon discovered at U.S. Naval Observatory, Flagstaff.
1991 Hubble Space Telescope clearly shows Pluto and Charon.

COMETS

There is always great excitement when a comet appears visible to the naked eye. A bright comet may have a graceful tail arcing through the sky. Comets are made of ice and small rocks. Two have been visited by the European space probe *Giotto*.

▷ The 1066 apparition of Halley's Comet was recorded in the Bayeux Tapestry, which was embroidered between 1067 and 1077. King Harold of England is being told that the comet is an evil omen.

▽ Comet Halley at the 1986 appearance, photographed with an ordinary camera. The blue tail is gas being released from the comet.

Comets are relics from the birth of our solar system. For billions of years they have spent their lives in the deep freeze of remote space. They are made of ices, such as frozen water, methane, ammonia, and carbon dioxide. Gritty dust, heavy stones, and lumps of metal are encased in this icy mixture. All these materials were present in the interstellar cloud from which the sun and planets formed. Comets are made from the stuff that got left behind. They are loosely assembled and can easily break up.

Journeying to the Sun

When a comet journeys to the Sun, changes take place. The Sun's rays first warm the icy snowball directly. By the time the comet is inside the orbit of Jupiter, solar heating starts to boil off the ices, and gas builds up around the speeding comet. A large, gassy cloud, named the coma, surrounds the comet. As the ice melts, a dusty crust builds up on the comet's surface. This is usually a very uneven layer, so the gas still being released inside starts to burst out in jets, almost as if the comet had miniature volcanoes.

Jet-propelled gas and dust streams from the comet, as it speeds toward the Sun. Fountains of material spewing from the comet can even change its orbit slightly. You see effects like this when a small balloon is blown up and then released: it darts around erratically as the air jets out of the neck.

As a comet races ahead along its orbit, the dust and gas get stranded, forming a comet tail. The radiation leaving the Sun pushes the tail away from the Sun. Although the nucleus of a comet head (the icy snowball) may be only a few miles in diameter, the tails can become up to 60 million miles long. Meanwhile, along the orbit of the comet, larger lumps of dust, stone, and metal flake away. These solid objects form meteor streams. After the comet has passed around the Sun, it cools as it moves farther away. The production of gas slows down, the jets stop, and the tail dwindles away.

Myths and history

The appearance of a new comet in the sky is unpredictable, but about every 10 years or so on average a bright one blazes forth and is visible in the night sky. For a few days at least, and perhaps for months, it remains visible. In some cases, a tail develops as well. If a comet passes close to Earth, the tail can stretch across a quarter or more of the sky. In addition to these unexpected comets, some of the comets that return again and again are also visible to the naked eye. Halley's Comet is the best known and brightest of these.

Comets may appear anywhere in the sky, can move in any direction across the sky, and change in brightness and shape. Because of their strange behavior, comets frightened people in the past.

In 467 B.C. a huge meteorite fell in Thrace at the same time a comet was visible for 75

▷ The artist Giotto probably used the 1301 apparition of Halley's Comet as the inspiration for his star of Bethlehem when he painted this biblical scene on a wall in the Scrovegni Chapel in Padua, Italy, in about 1304.

days. As a result, the Greek thinkers believed that hot stones were part of the sky. Aristotle, the most famous of the ancient Greeks who wrote about the natural world, was born in Thrace and worked in Athens in the 4th century B.C. In his textbook about the weather, he said that comets caused strong winds and drought. His ideas were generally accepted by educated people for almost 2,000 years. After the assassination of Julius Caesar (44 B.C.), a bright comet was seen in all parts of the Roman Empire for a week. People thought this was Caesar's spirit, joining the gods in the heavens.

Another famous example of a comet being linked with disaster was in 1066. The Norman invasion of southern England coincided with an appearance of Halley's Comet. This happened again, at the fall of Constantinople in 1456. Throughout the Middle Ages, scholars and writers repeatedly said comets brought nothing but trouble.

In 1577 a comet appeared that was so bright it could be seen through the clouds. The Danish astronomer Tycho Brahe (1546–1601) observed it, and showed that it must be traveling through space far beyond the Moon. This completely disproved the scary theories of Aristotle and the others who claimed that comets were dangerous weather effects.

Newton and Halley

Isaac Newton proved mathematically that all comets are orbiting the Sun, and are controlled by the Sun's gravity. He also showed that the comets are on stretched-out orbits. They are visible in our sky only when they swing through the solar system and get near the Sun.

Edmond Halley was a friend of Newton. Halley worked out the orbits of comets using Newton's methods. He found that the comet he observed in 1682 had the same orbit as the one seen in 1607. In a thrilling piece of scientific detective work, he next tracked back to 1531: the comet of that year was on the same orbit! Clearly these three apparitions were of the same object. The comet returns every 76 years or so, and Halley said it would come again in 1758, but he did not live to see it.

Predictable comets

Halley's Comet did return in the winter of 1758, exactly as Halley predicted. From that time it was known as Halley's Comet. It was the first comet apparition to have been predicted in advance. In fact, it has been seen at every return to the Sun since 240 B.C., and it may have been observed by Chinese skywatchers as long ago as 1059 B.C. Its most recent appearance was in 1985–86. Its next return to the inner solar system will be in 2061, and it will be observable from 2060 to 2062.

The predictable, or periodic, comets take between 3 and about 200 years to orbit the sun, traveling on orbits that stay within our planetary system. The comet with the shortest orbital period is Encke's Comet, first seen from Paris in 1786. It was rediscovered in 1795, 1805, and 1818. J. F. Encke (1791–1865) computed the orbit of the 1818 comet and linked it with the previous apparitions. He predicted a reappearance in 1822 and it was found again as a result. Its orbital period around the Sun is only 40 months.

New comets

The unpredictable comets come from far beyond the realm of the planets. Although they are members of the solar system, they take millions of years to come from the depths of the solar system. When one arrives, it whizzes past the Sun in just a few weeks, before returning to more or less the place from which it came. When close to the Sun, the comet brightens up as the Sun's rays boil off some of its gas, and it will often have a tail.

Fewer than 1,000 comets are known. Many of the new ones are found by amateur astronomers who search for them carefully, using small telescopes. Comet seeking requires a lot of skill and patience. First you have to get to know the locations of all the fuzzy objects, such as galaxies and nebulae. Only then can a new arrival be spotted without causing a false alarm. Most searchers get up really early, while it is still dark, and sweep the dawn sky. Very active comet-spotters can expect to find a new one every couple of years or so.

Professional astronomers sometimes find a new comet by accident, when it shows up on a photograph that has been taken for another purpose. One space telescope, the Infrared Astronomy Satellite (IRAS), discovered six comets by detecting the heat radiation from the warm dust around the comets.

Suspected new comets are reported immediately to an office at the Smithsonian Astrophysical Observatory in Cambridge, Massachusetts. As soon as other astronomers have confirmed the discovery, the information is sent through computer networks worldwide to everyone who is interested in observing comets.

Naming comets

A new comet is named after its discoverer. If several people report a discovery independently, only the first three to contact the comet office in Cambridge, Massachusetts, get their names attached

to it. A few comets—Halley and Lexell, for example—were named after the mathematicians who computed their orbits. Those found by the IRAS orbiting observatory are named after it. Short-period comets in elliptical orbits around the Sun have "P/" put in front of their names. The formal name for Halley's Comet is Comet P/Halley.

The comet cloud

When a comet returns to the Sun, it loses material through heating and the creation of the tail. Eventually, the comet will burn

out. In fact, Halley's Comet is dimmer now than it was in the past. The periodic comets will all be destroyed in less than a million years. This is a short time compared to the age of the solar system. New comets must be coming from somewhere to replace the ones that have burned out.

When completely new comets are discovered, they appear to have traveled at least 50,000 astronomical units (about 4.5 million million miles) to reach the inner solar system. Some astronomers suggest that a shell of comets, stretching one-quarter of the way to the nearest star, surrounds the sun. This comet cloud,

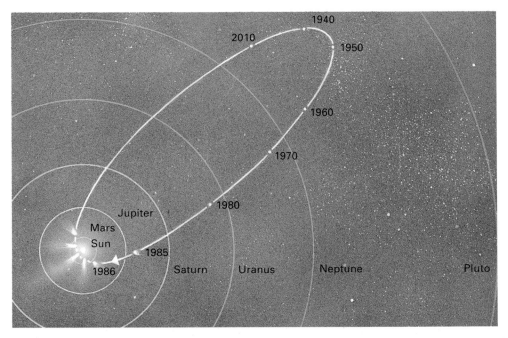

△ Comets that return to be seen time and again travel through the solar system on elliptical orbits. The orbit of Halley's Comet, shown here, is very elongated.

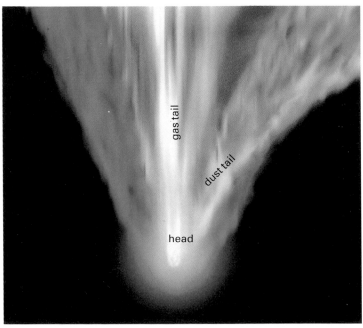

◁ The head of a comet contains the nucleus, made of ice, dust, and rock. The nucleus is surrounded by a cloud of dust and gas, called the coma. Many comets develop two tails. The gas tail always points away from the Sun, while the dust tail usually curves around.

which has not been seen in any telescope, marks the edge of the solar system. Other astronomers have suggested that the Kuiper Belt, which is much closer in, is a source of comets.

Comets get shaken out of the distant cloud or the Kuiper Belt as the solar system journeys around the Milky Way Galaxy. Effects of the gravitational force of the Galaxy cause a trickle of comets to leave the cloud and fall toward the Sun. The comet cloud contains 100 billion comets.

When a new comet reaches the realms of the planets, it is in danger if it passes close to Jupiter. The effects of the gravitational pull of this massive planet can swing a comet out of its original orbit and leave it trapped in a much smaller elliptical orbit: this is the way in which the family of periodic comets gets replenished with new members.

A visit to a comet

Three comets have been visited by spacecraft. In 1978 a space probe passed through the tail of Comet Giacobini–Zinner. For the 1986 return of Halley's Comet, a flotilla of craft were launched: two Russian spacecraft went through the gaseous tail, and a Japanese spacecraft observed a hydrogen cloud around Halley. The European spacecraft *Giotto* skimmed right by the nucleus itself, obtaining photographs that showed the coal-black surface of the comet and fantastic jets of gases bursting from inside. In 1992 *Giotto* flew by a second comet, Grigg-Skjellerup. In the future, space scientists plan a mission to a periodic comet that would enable samples of the nucleus to be put on board a spacecraft and returned to Earth. The European Space Agency plans to send a probe to a periodic comet, arriving in about 2011.

◁ The spacecraft *Giotto* obtained this close-up of the nucleus of Halley's Comet from 1,200 miles away. The nucleus is very dark and about 9 miles long by 5 miles wide. Jets of gas and dust are streaming from holes on its surface.

A COMET HITS JUPITER

Comet Shoemaker-Levy 9, discovered in 1993, was extremely unusual. The gravitational pull of the giant planet Jupiter trapped this comet into an elongated orbit around Jupiter. On a close approach to Jupiter in 1992, tidal forces tore the comet into more than 20 separate pieces, each of which continued on its own orbit. The broken comet looked like a string of pearls. Toward the end of July 1994, the fragments crashed head on into Jupiter, causing spectacular changes in the atmosphere. This was the first time astronomers could witness the effects of an impact between a comet and a planet.

SOME REMARKABLE COMETS

Comet Arend-Roland (1957) was one of the more prominent 20th-century comets. It developed a spike pointing toward the Sun.

Comet Biela (1772) split in two during the 19th century; although it was periodic, it subsequently vanished in 1852.

The Daylight Comet (1910) could be seen in broad daylight. It appeared a few months before the 1910 return of Halley's Comet.

Comet Donati (1858) was a brilliant comet that developed a curved dust tail and two thin gas tails.

Comet Encke (discovered in 1786) has a very short period of 40 months, and has been observed to orbit the Sun more times than any other comet.

Comet Halley is the longest observed comet.

Comet Ikeya-Seki (1965) was the most recent comet that was very easy to see.

Comet Kohoutek (1973) was discovered while right out at the orbit of Jupiter. For a time it seemed that it would become the comet of the century, but it failed to blaze out as it neared the Sun.

ASTEROIDS

Between the orbits of Mars and Jupiter are many thousands of small objects made of rock. The largest is almost 600 miles across, but most are much smaller. They are rocks left over from when the planets formed.

On January 1, 1801, the Italian astronomer Giuseppe Piazzi (1746–1826) discovered a tiny object orbiting between Mars and Jupiter. At the time he was preparing a chart of the stars, but he noticed one point of light that had moved to a new place since he had last mapped that part of the sky. This moving "star" was the first asteroid, which he named Ceres. Its diameter is 567 miles, and it is by far the largest of the asteroids. Still, it is quite small compared to all the planets and many of their moons. Our Moon is nearly four times larger in radius. There are several moons in the solar system that are larger than Ceres. For that reason, Ceres, and the other objects between Mars and Jupiter, are known as minor planets.

Many thousands of minor planets, or asteroids, are now known. Almost all of them take three to six years to orbit the Sun. They are swarming between Mars and Jupiter, in a zone called the asteroid belt. Another belt was discovered beyond Neptune in 1993.

The biggest asteroids are from 10 to 50 miles across. Ceres and a handful of other asteroids are quite exceptional because they are as big as some planetary moons. Most known asteroids are smaller than 10 miles. Astronomers think that millions of boulders, stones, and grains of sand are cruising through the asteroid belt.

The Trojans

Asteroids can get trapped if they stray too close to Jupiter. There are two families of asteroids that march around the solar system in front of and behind Jupiter: they are named the Trojans. Occasionally, one of them tumbles right into Jupiter's gravitational pull and becomes a tiny satellite of Jupiter.

Chaos in the solar system

Mathematicians have discovered that some asteroids can follow all kinds of bizarre tracks. The big planets in the solar

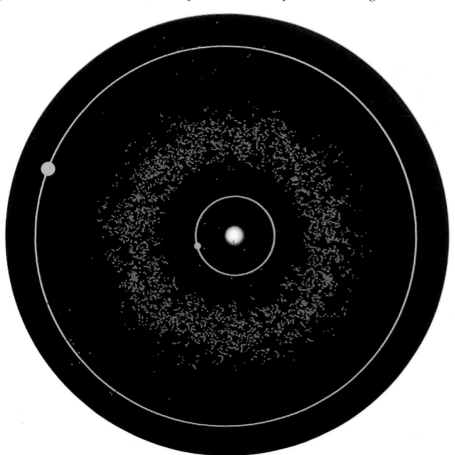

◁ The actual positions of known asteroids on April 8, 1992, with the orbits of Earth and Jupiter. A number of asteroids are in orbits crossing Earth's.

ASTEROID FACT FILE

Largest: Ceres, 567 miles diameter

Smallest known: 1991 BA, 30 feet diameter

Darkest: 95 Arethusa is as black as coal.

Closest approach to Earth: 1991 BA, 100,000 miles

Closest approach to the Sun: Icarus and Phaeton. Both get nearer than planet Mercury.

Farthest asteroid: this may be Chiron, discovered in 1977. Its orbit is entirely between Saturn and Uranus, and its diameter is about 120 miles. It may be a dead comet or an escaped satellite of Saturn.

◁ The spacecraft *Galileo* obtained this image of the asteroid Ida in 1993 from a distance of about 1,500 miles. Ida is about 32 miles long. There are many small craters caused by meteorite impacts.

system tug them this way and that. Their orbits can suddenly go off track, and then an asteroid can be hurled toward the Sun. Astronomers have already found more than 1,000 asteroids with orbits that cross Earth's annual path around the Sun. The tiniest satellites within the solar system could be asteroids that were trapped by their parent planets.

Meeting an asteroid

Scientists using telescopes on Earth cannot see surface markings on asteroids because asteroids are too small and too far away. However, the spacecraft *Galileo* dropped in on two asteroids, Gaspra and Ida, while journeying to the giant planet Jupiter. Gaspra is shaped like a potato and has a few small craters that were probably made by meteorite impacts long ago. This rocky world has clearly suffered massive damage in the past. Scientists believe that two huge chunks were ripped away in collisions with other asteroids. Most asteroids have irregular shapes. Ida has a tiny moonlet accompanying it, which is about one mile across.

By studying asteroids, astronomers hope to learn more about the material from which the planets are formed.

Near-Earth objects

In the recent past (astronomically speaking), comets and asteroids have slammed into Earth. The Meteor Crater in Arizona is only 50,000 years old. The Tunguska event in Siberia occurred in 1908. This seems to have been an explosion of an asteroid or worn-out comet in the atmosphere. The blast-wave flattened trees in an area more than 60 miles across.

Impacts like Tunguska would be catastrophic in a populated area today. In 1991 a small asteroid skimmed just 100,000 miles from Earth. Several known asteroids will pass within 500,000 miles of Earth in the near future. Astronomers are making searches for near-Earth objects (NEOs) of this kind so that they can detect in advance any that might come dangerously close. It might be possible to send a nuclear bomb into space to nudge to one side any really dangerous asteroid.

Already 150 out of the 10,000 known asteroids have come very close to Earth. Some of these are simply large rocks that would do no harm. The Spacewatch Telescope, designed to search for small asteroids, can detect objects smaller than 100 feet, and some are less than 30 feet in size. A rock this small smashes into Earth once in a lifetime and is not dangerous.

A few near-Earth asteroids are targets for space missions between now and 2007. They may contain minerals that are rare and valuable on Earth.

So far as Earth is concerned, an impact with an asteroid a little larger than half a mile happens once every 100,000 years, on average. Such a collision would be catastrophic for the whole world.

The dinosaur age came to an end 65 million years ago, at the same time that a small asteroid smashed into the Yucatán Peninsula, Mexico. Debris was blasted high into the atmosphere and fires raged all over Earth. The fireball from this impact triggered a global environmental crisis. The resulting climatic change wiped out two-thirds of the species then on Earth. No animals weighing more than 60 pounds survived. Small mammals the size of mice probably survived because they were burrowing animals. The mammals of today are thought to be descended from these small creatures.

Of all the objects that astronomers study, only asteroids and comets can affect Earth in a way that could cause disaster. However, the chance of such a thing happening to us is very small indeed. Many of the world's people are much more at risk from earthquakes and volcanoes, disease and famine.

METEORITES

Meteorites are stones that fall from the sky. Mostly, they are left over from the formation of the solar system, but a few have come from the Moon and even Mars.

Between the planets there is a surprising amount of cosmic debris. Much of it is material left behind when the planets formed, but some is more recent, such as the dusty trails left by comets. To describe this material astronomers use three similar words: meteoroid, meteor, and meteorite.

A meteoroid is a piece of rock or dust out in space. Earth is constantly bombarded by objects ranging in size from specks of dust to rocks weighing several pounds. These enter the atmosphere at speeds of about 37,000 miles per hour, or more. Friction with the air heats the particles until they glow red-hot. A meteor is the visible trail in the sky left by an object that burns up as it enters the atmosphere. These trails are also called shooting stars. A meteoroid that reaches the ground is a meteorite. Individual meteorites are often named after the place where they fell.

Meteor streams and showers

During its annual motion around the sun, the earth sweeps up about 1,000 tons of cosmic rock and dust. Much of this material is orbiting the solar system in streams, where comets have left a trail of rubbish as they hurtle through the solar system. When the Earth crashes through one of these streams, a meteor shower may be seen. Streaks of light, caused as dust specks burn up in the atmosphere, appear to radiate from a single point in the sky. Meteor showers are fairly predictable, because the Earth tends to cut through a stream on more or less the same date each year.

Meteorites

Rocks that survive the fiery journey through the atmosphere are not all that common. About 200 tons are estimated to

△ Meteor trails on a time-exposure photograph. The many parallel, broken trails were made by stars gradually moving across the sky during the exposure. The meteor trails were bright for just a few seconds.

▽ An iron meteorite that has been cut through. The cut surface has been treated to show the pattern of iron and nickel crystals inside the meteorite.

◁ Meteor Crater in Arizona is 1 mile wide and 600 feet deep.

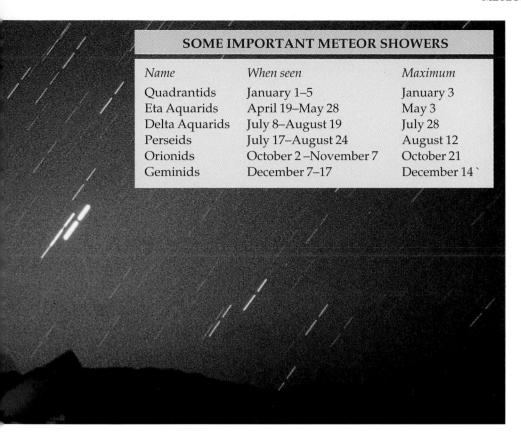

SOME IMPORTANT METEOR SHOWERS

Name	When seen	Maximum
Quadrantids	January 1–5	January 3
Eta Aquarids	April 19–May 28	May 3
Delta Aquarids	July 8–August 19	July 28
Perseids	July 17–August 24	August 12
Orionids	October 2 –November 7	October 21
Geminids	December 7–17	December 14 `

SEE A METEOR SHOWER

Very occasionally, an exceptional meteor shower may give tens or even hundreds of meteors per minute, but around 50 per hour is more typical. If you would like to see a meteor shower, the most reliable one is the Perseids, which is best during August 10–14 each year. Under the most favorable conditions, there may be 75 meteors per hour. Meteors are easiest to see if there is no moonlight, and if you wait until midnight or later. You can look almost anywhere in the sky. Meteors in a shower appear to come from one area of sky. The Perseids radiate out from the constellation Perseus. Allow at least 20 minutes of total blackness for your eyes to adjust fully to the dark. The table at left gives the dates of showers at other times of the year.

hit Earth's surface, almost all as very small dust grains. Only 20 or so meteorites are found each year from new falls. Radioactivity in meteorites shows they were formed 4.6 billion years ago, as part of the solar system. Because they are samples of the primitive material from the early solar system, they are much sought after by planetary scientists.

There are three main kinds of meteorites: ones made largely of iron; stony-irons, with one-third to one-half of their volume as metal; and finally, the stones, which may contain small amounts of metal. The iron meteorites are easiest to recognize because they are dense and durable. The stony meteorites are interesting because they have never been intenselyheated (apart from during their brief passage through the atmosphere). This means they have not changed very much since their formation. As a result, their chemical makeup is probably similar to that of the early solar system.

There is no record of any human death from a meteorite, although there have been near misses. One meteorite landed on August 31, 1991, less than 13 feet from two boys in Noblesville, Indiana, making a small hole 2 inches deep and 4 inches across. The same year, another whizzed down close to a man working in his garden in Peterborough, England. In October 1992 a large meteorite destroyed an empty car in New York State.

Large meteorites have left substantial craters. The one in Arizona is the best preserved, because the dry desert climate has prevented erosion and weathering since its formation about 50,000 years ago. However, it is just one out of 150 terrestrial meteorite craters, many of which are much larger. One of the largest, Manicougan, in Quebec, Canada, is 200 million years old, and 60 miles in diameter.

The Antarctic ice-sheet is now the main source of new meteorites for analysis. Thousands have been recovered already. After being buried in snow and ice for up to a million years, they have been exposed in regions where the ice cap is being scoured away by strong winds, and are found lying on the surface. The dry and stony deserts in western Australia and Namibia are also important sources of ancient meteorites.

▷ Collecting a meteorite from the Antarctic ice-sheet.

Rocks from the Moon and Mars

The meteorites from Antarctica have included rare and exotic specimens. Some are similar to rocks returned from the Moon. It is thought that these small stones were ejected from the Moon's weak gravity by an asteroid impact about 100,000 years ago. Even rarer are a handful of meteorites that may have come to us from Mars. These seem to be about 1.3 billion years old, and were launched into space during a huge impact. Geologists recently extracted a single drop of water from Martian meteorites, a sample of the oceans that once existed on the now-dry planet.

THE SUN AND THE STARS

Why is the Sun so hot? Where can you see a star factory in the sky? What is the difference between a white dwarf, a black hole, and a red giant? Stars are giant balls of glowing gas. Deep inside they generate nuclear energy. Stars are born in the gas clouds of space. In the end, they either explode or just fade away.

Interstellar Matter

▷ The whole of the Milky Way galaxy, imaged by the Cosmic Background Explorer satellite. This image mainly shows the heat radiation from dust in the Milky Way.

Deep space between the stars is not empty. Gigantic clouds and swirling masses of gas and dust form beautiful glowing clouds of matter. These are called nebulae (from the Latin word for clouds), and many are the birthplaces of new stars. Inside the Orion Nebula, new stars are forming right now.

To see the dust clouds of the Milky Way with the naked eye, you need a night when the Moon is not up, and you must be well outside the bright lights of towns and cities. Then you can pick out a ribbon of faint light across the sky, about as wide as your outstretched hand. The Southern Hemisphere is the best place to see the Milky Way, but it is not hard to see it during summer nights in the Northern Hemisphere. Among the wisps of light are gaps and holes, which show up well on photographs.

For a long time astronomers thought these dark patches in the Milky Way were tunnels through the stars. We now know this idea is completely wrong. The regions with few stars are really clouds of smoke and dust. Soot and grains of sand are sitting out there in deep space, and they block our view of the stars in the Milky Way. How much of this fog is there in outer space? It blocks so much light that if we could blow it all away you would be able to read a book at night by the bright light of the Milky Way alone.

▽ The plane of the central part of the Milky Way, imaged with the Infrared Astronomical Satellite, showing the gas and dust concentrated into a thin band.

The effects of dusty space

In our own atmosphere, the setting Sun looks red because dust in the air deflects blue light more than red light. So most of the red light gets through the haze, but not much blue light. In much the same way, the smog in space not only makes the stars dimmer but makes their light redder, too. Toward the center of our Galaxy, in the constellation Sagittarius, there is so much dust that nothing gets through to us from the center. To find out what is going on at the heart of the Milky Way, astronomers must use radio and infrared telescopes to penetrate the haze.

Specks of dirt in space are heated up by starlight, particularly in the vicinity of very hot stars. Using special infrared telescopes, astronomers can see the heat radiated by dust grains, and so these telescopes let us peer into dust clouds. When part of a gas or dust cloud starts to shrink as a result of gravitational forces, it has to get rid of energy released by the collapse. This energy is seen as infrared radiation.

Stardust

The dust in the Milky Way was made by the stars. The outer layers of giant stars get wafted away into space. Old stars explode and scatter oxygen, carbon, and iron atoms into space. Silicon and iron form very tiny crystals that then drift into space where they pick up a coating of oxygen, carbon, and nitrogen. These little specks are miniature chemical factories. On the surfaces of the dust grains, atoms, such as carbon and oxygen, get stuck together to make molecules—carbon monoxide, for example.

Hello! Hydrogen calling Earth!

The most common substance between the stars, and in the whole universe, is hydrogen. Radio astronomers listen to the hiss produced by this gas all through our Galaxy. A hydrogen atom has just one electron. Sometimes the electron flips over, and as it does so a blip of radio energy is sent out. Each blip is feeble, but there is so much hydrogen in space that astronomers can pick up the combined effect from all the hydrogen as radiation at a wavelength of 21 centimeters (8.25 inches). Maps of the hydrogen in the Milky Way reveal the beautiful spiral shape of our Galaxy, with much of the hydrogen squeezed into the coiling arms.

Clouds of hydrogen orbit the Galaxy in the same way that planets orbit the Sun. The speed of a hydrogen cloud depends on how far it is from the center of our Galaxy. From the speeds of hydrogen clouds, we can find the overall size and shape of the Galaxy.

Emission nebulae

Interstellar clouds are made mainly of hydrogen. In deep space they are too cold to send out any light. Sometimes a cloud of hydrogen contains a very hot star. Then a beautiful emission nebula is seen as a cloud of glowing gas. The hot star heats up the hydrogen cloud until it glows with pinkish light. There is a huge pink emission nebula in the Large Magellanic Cloud.

Absorption nebulae

An interstellar cloud may be too cold to emit any light. However, a cold cloud can absorb the light from bright objects (such as stars) behind it. Then it may be seen as a silhouette against the background light. The Coalsack, a dark patch in the southern Milky Way, is an absorption nebula that is visible to the naked eye.

Reflection nebulae

A cool cloud in space can sometimes be seen because dust in the cloud reflects the light from nearby stars. Dust makes a wispy reflection nebula around the brightest stars in the Pleiades cluster. Reflection nebulae look blue in photographs.

The interstellar medium

The matter between the stars is known as the interstellar medium. Most of it is concentrated into the spiral arms of the Milky Way. The temperature varies from just a few degrees above absolute zero in the coldest dust clouds up to a million degrees in the hottest gas clouds.

If you went out in space to a spiral arm of the Galaxy, you would find one to two atoms of gas per cubic inch. In about a cubic mile, there would be a thousand dust grains. So, on average, the interstellar medium is very thinly spread out. However, inside the thick clouds it can get 1,000 times denser than this. Even so, a dense cloud has only a few hundred atoms in a cubic inch. The reason why we can see interstellar matter even though it

▽ ▷ The Trifid Nebula (below left) shows clouds in emission (red light) and reflection (blue light). The magnificent Horsehead Nebula in Orion (below center) is a dark absorption nebula seen against background light. Rho Ophiuchi (right) also shows both emission and reflection clouds.

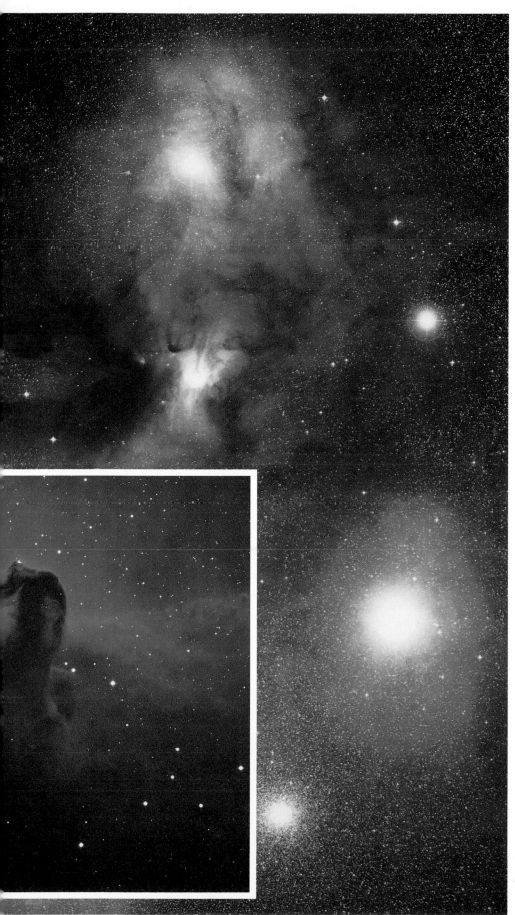

is so spread out is because we are looking through a great thickness. The interstellar matter makes up about 5 to 10 percent of the visible matter in a typical spiral galaxy.

Our solar system is in a region where the density of the interstellar medium is unusually low. This is called the Local Bubble and it extends out to around 300 light-years. Perhaps some process has caused most of the matter we might expect to find near the Sun to be blown away. One idea is that, long ago, several huge stars may have exploded in our vicinity. The blast waves from the explosions could have swept the interstellar gas farther away in space.

Giant molecular clouds

The most massive objects in the Milky Way are giant molecular clouds. They are up to a million times more massive than the Sun. The Orion Nebula is just part of a giant molecular cloud, about 500 times more massive than our Sun. Within the hidden depths of the black clouds, astronomers have found an amazing selection of molecules. The stuff in space includes water, ammonia, and alcohol. There is formic acid, which is found in stinging ants, and hydrogen cyanide. Many of the substances are of the kind that chemists call organic molecules, because they contain carbon.

The chemistry of these marvelous clouds is really quite simple. We can imagine the different atoms as parts of a construction kit. Carbon, hydrogen, oxygen, nitrogen, and other atoms are put together in different ways to make various molecules that do not break up because the cloud is very cold. Very simple units can be linked together to form molecules of amino acids and proteins. On Earth, these substances, which exist naturally in space, are linked together to make the giant molecules of plant and animal life.

Cosmic rays

Very high-speed atomic particles, such as nuclei and electrons, zip through space at nearly the speed of light. These are cosmic rays. So far astronomers have not been able to say for certain how these cosmic rays are created. Probably they come from within our Galaxy, perhaps being flung into space by exploding stars.

THE SUN

The Sun is a normal star and is about 5 billion years old. The surface temperature is about 10,000°F, but at the center it is 25 million °F. In the Sun's core, hydrogen turns into helium, and this releases energy. On the surface there are sunspots and solar flares, and great explosions can be seen.

We are living right next to a real star, the Sun, which is at the center of our solar system. It gives the Earth heat and light, both of which sustain life. Plants, for example, use sunlight as a source of energy for growth. Fossil fuels, such as coal, are actually a form of the Sun's energy that has been stored away, because the carbon they contain was made by plants long ago.

For astronomers the Sun is a special star because we are so close to it. We live 93 million miles away. Driving an ordinary car that far would take nearly 200 years, so it is a long way even to our home star. A space probe traveling along a straight path would take many months to get to the Sun. Light, which travels faster than anything else, zips along in just over eight minutes from Sun to Earth. Proxima Centauri, the next nearest star, is more than a quarter of a million times farther away than the Sun.

We know far more about the Sun than any other star simply because it is so close to us. Some large observatories have telescopes that are used only to study the Sun. Astronomers want to know how the Sun works and how it affects the Earth. From this we can understand how most of the normal stars work.

Some scientists have suggested that any changes in the Sun's output of energy may alter the climate here on the Earth. Solar astronomy is important, therefore, both for the study of the stars and for seeing how our environment may be directly affected by the sun in the future.

THE MESSAGE IN SUNLIGHT

The spectrum of light from the photosphere carries a great deal of information. In 1814 the German physicist Joseph Fraunhofer (1787–1826) found hundreds of dark lines crossing the spectrum. He described and listed 700 of them. We now know that these spectral lines, of which there are many thousands, are caused by different chemical elements in the solar atmosphere located in a cool layer above the photosphere. Iron, for example, is responsible for many of them and sodium contributes a pair of dark lines in the yellow part of the spectrum. By studying the lines in the spectrum, astronomers can find out what elements the Sun contains and in what proportions.

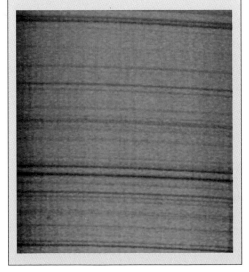

◁ The modern solar observatory at Teide, Tenerife, Canary Islands, is located high on an extinct volcano. This is Europe's main observatory for studying the Sun. It is situated in a dry and relatively cloudless climate. The main solar observatory in the United States is located at Kitt Peak, Arizona.

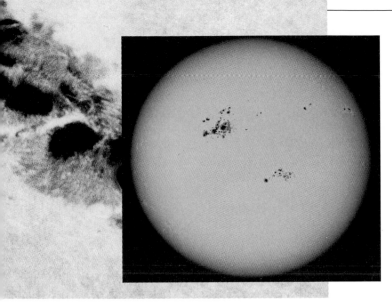

◁ A giant sunspot group (June 12, 1991) (far left) and two giant groups of sunspots (October 29, 1991). The average temperature within sunspots is about 8,000°F, and they only appear dark because they are cooler than the rest of the Sun's photosphere. Sunspots are generally found on either side of the Sun's equator, but not on the equator itself.

Sunspots

At the Sun's surface layer, where its energy finally bursts forth as light, astronomers see many different kinds of activity. Sunspots are an obvious sign of this activity. They are areas on the Sun's disk that are cooler and less bright compared to the brilliance of the photosphere.

It is sometimes possible to see very large spots just as the Sun is setting, and this is how Chinese astronomers made observations 2,000 years ago. The early astronomers believed that the spots were a phenomenon in our own atmosphere, but this idea was disproved in the 17th century by Galileo. He used his telescope to follow sunspots in 1610 and made many important discoveries. For example, Galileo found that spots appear and disappear, and that they change in size. By following their motion across the Sun's disk, he showed that the Sun rotates. He also observed that spots change in shape as they near the edge of the visible disk.

A sunspot has a dark central region known as the umbra. This is surrounded by a less dark penumbra. Giant sunspots are regions where strong magnetism inside the Sun wriggles through the surface layers. Large spots are bigger than Earth and they may last for a couple of months.

The surface

The Sun is a fiery ball of gas, about 109 times larger in diameter than the Earth. The Earth's volume would fit inside the Sun more than 1 million times. The yellow light of the Sun comes from a layer in its atmosphere about 300 miles thick called the photosphere. Below this layer lies the solar interior and above it parts of the outer atmosphere that are transparent. Practically all the Sun's energy, including the heat and light that is falling on Earth, comes from the photosphere, but was originally made deep inside the Sun.

The temperature of the photosphere is about 10,000°F. One way of estimating this temperature is to determine how hot the Sun must be in order for it to emit all the energy it gives out.

The Sun's surface is bubbly. This froth is called the solar granulation, and it can only be seen with solar telescopes. The bubbling is similar to that seen when milk or gravy boils. Convection in the Sun's atmosphere is carrying heat energy from the lower levels to the photosphere, causing the frothy texture.

In the 1960s astronomers discovered that the photosphere's upper layer moves in and out once every five minutes or so. So the Sun is vibrating, like a ringing bell. By studying these vibrations, astronomers hope to discover what the inside of the Sun is like.

◁ The bubbly structure of the Sun's photosphere, known as granulation.

HOW TO SEE SUNSPOTS

To observe sunspots, a small telescope can be set up to produce a solar image by projection. To avoid looking straight at the Sun, which is dangerous, the eyepiece of the telescope is set a little farther out than the normal viewing position. By trial and error, the telescope is pointed at the Sun. With practice, a sharp image can be focused onto a card held four to eight inches beyond the eyepiece.

Clip a sheet of paper to the card on which the image is projected and use a pencil to mark the positions of the sunspots. By repeating this experiment for a few days you can see that the spots move from east to west as the sun rotates.

Action on the Sun

The Sun does not rotate like a solid body such as Earth. Instead, different parts rotate at different rates. The equator turns fastest, taking 25 days or so for one rotation. Rotation is slower away from the equator and in the polar regions takes around 35 days. The different rates of rotation are only possible because the Sun is a ball of gas. One effect of this kind of rotation is to wind up the Sun's magnetic field, which in turn increases solar activity.

Sunspots are just one example of solar activity. The "weather" in the Sun's atmosphere is very different from that on Earth. Magnetic storms and explosions known as flares blow up without warning on the solar surface. In some ways they resemble thunderstorms on Earth, because electrical energy is released. However, the energy in the Sun's giant electrical sparks is far greater. Earth is affected by solar storms, and for that reason astronomers keep the Sun under constant watch. The Sun's flares blast electric particles into space, which have wonderful effects on our atmosphere.

The aurora

If clouds of electric particles from solar flares reach Earth, they create marvelous curtains of shimmering light in the sky, seen in polar regions as the aurora. The dancing light of the aurora is very beautiful, but the giant solar explosions have their dangers as well. In a few seconds they blast out more energy than all the electrical power stations in the world have ever made. A giant solar storm in 1987 cost $100 million in damage to electricity supplies in North America. Electrical currents from the Sun forced power stations to shut down and destroyed equipment. Flares are dangerous to astronauts, too, who must not space walk when one is in progress. The high-energy particles from the flares can damage the human body.

The aurora is unpredictable and therefore difficult to observe. It can take the form of arcs, rays, and curtains of light in the sky, and no two displays are ever the same. A moonless night is essential and it is much easier to see in far northern or southern latitudes, such as Scotland, Nova

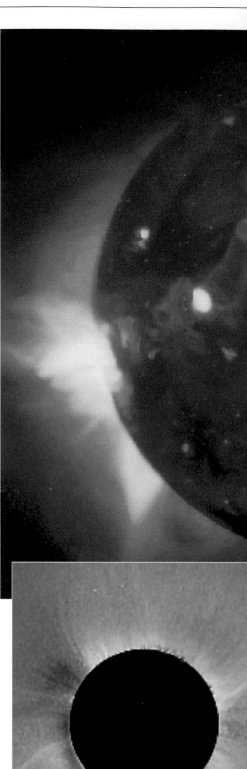

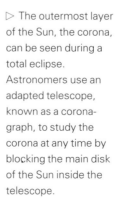

◁ A huge solar prominence leaps millions of miles into space. Most of the material will fall back to the Sun, but some is propelled through the solar system.

▷ The outermost layer of the Sun, the corona, can be seen during a total eclipse. Astronomers use an adapted telescope, known as a coronagraph, to study the corona at any time by blocking the main disk of the Sun inside the telescope.

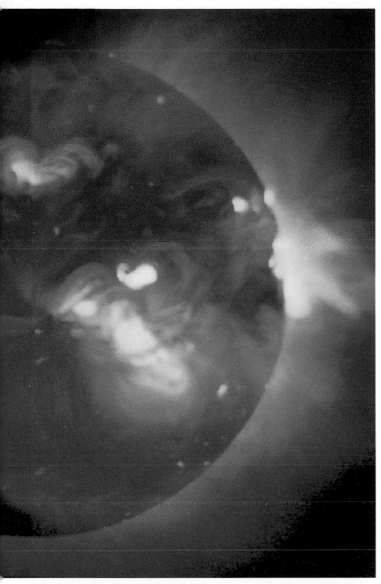

◁ The Sun imaged by the *Yohkoh* X-ray astronomy satellite. X-rays show the hottest parts of the solar atmosphere, where the temperature reaches up to 1,800,000°F.

◁ Electrically charged particles from the Sun cause the aurora emission in Earth's upper atmosphere. An aurora display is most likely when sunspot activity reaches a maximum.

Scotia, and Alaska (Northern Hemisphere), or the South Island of New Zealand. Auroral displays are not likely to be seen when there are very few sunspots.

The solar cycle

The number of sunspots that can be seen on the Sun varies with time. In 1989–90 there were a great many, for this was the peak of a cycle of activity. On average, the sunspots come up to a maximum about every 11 years. They will next be at their greatest density in about the year 2000 or 2001. In the mid-1990s sunspots will be relatively scarce.

The cycle of sunspot activity appears to be linked to the climate on Earth. Some trees, for example, show variations in the thickness of their growth rings running through an 11-year cycle. Between 1650 and 1715 there were hardly any sunspots, as if the solar cycle had vanished altogether. This corresponded to a period of intensely cold weather in Europe.

To see if the 11-year solar cycle affects our climate, an instrument on a satellite measured the amount of energy produced by the Sun in the period from 1980 to 1989. Each time large sunspots appeared on the Sun, the amount of energy radiated by the Sun fell. A new series of spacecraft observations is taking place throughout the 1990s. Scientists expect these measurements to show whether the variations on the Sun are producing long-term effects on Earth, such as contributing to global warming.

Outer layers of the sun

Solar eclipses enable astronomers to see the layers of the sun's atmosphere above the photosphere. A ring of pinkish light comes from the chromosphere, where the temperature is about 27,000°F. During the total eclipse, a faint white halo, the corona, is visible for a few minutes. This actually extends to several times the radius of the sun. Its highest temperature, near to the sun, is 3,500,000°F.

The intensely hot corona sends out very little light but a great deal of X-ray energy. Astronomers use X-ray telescopes on orbiting satellites to view the X-ray sun. Computer graphics are used to make color images of the regions emitting X-rays.

◁ Earth is surrounded by a magnetic cage, which deflects most of the electrically charged particles sent out by the active Sun.

▽ A cross-section of the Sun. The helium core is about one-quarter the radius of the whole Sun; it contains much of the Sun's mass. The energy made in the core slowly travels through the thick hydrogen layer, to heat the visible photosphere.

This is how we know that the bright regions of the corona have temperatures of up to 1,800,000°F. The cooler regions in the corona appear as dark holes through which particles such as electrons can stream into space.

The solar wind

The outer layers of the Sun's corona are steadily blowing away right through the solar system in a breeze known as the solar wind. It takes about 10 days for the particles to travel as far as Earth. The *Voyager* spacecraft found they could detect the solar wind even beyond the orbit of Pluto. The solar wind forces the gas in the tails of comets to point away from the Sun.

Earth's magnetic cage

Earth has a magnetic field, which deflects most of the solar wind and keeps the particles from striking our planet directly. In effect, Earth's magnetism makes an invisible cage around which the solar wind flows, like a river around an island. Other planets with magnetic fields, such as Mercury and Jupiter, also have unseen barriers to the solar wind. In the case of Earth, some electrically charged particles leak through.

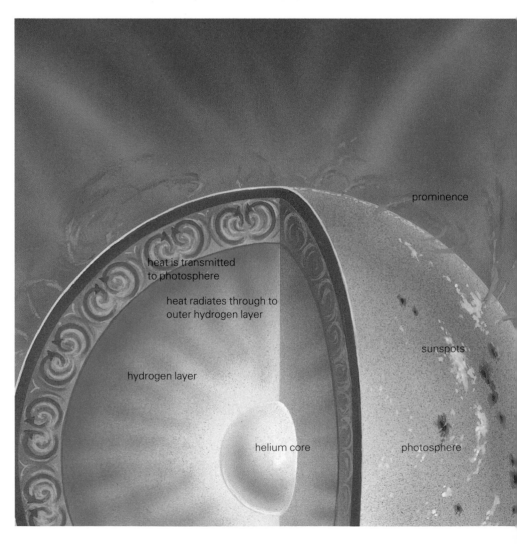

prominence

heat is transmitted to photosphere

heat radiates through to outer hydrogen layer

hydrogen layer

sunspots

helium core

photosphere

MAKING THE SUN'S ENERGY

The Sun is mainly made of hydrogen, the lightest chemical element, and helium, the second lightest. Deep inside the sun, the very high temperature prevents ordinary atoms from forming. Instead, atomic nuclei and electrons are all mixed up. (Chemical elements, atoms, and the particles in them are explained on pages 20–21.) The hydrogen nuclei are single protons. Helium nuclei are clusters of two protons and two neutrons.

At the incredible temperature at the center of the Sun, atomic particles move around at very high speeds and often collide. Usually nothing happens. Sometimes, though, two protons collide with such force that they stick together and change into a proton-neutron pair (1). Then they emit two other particles: a tiny neutrino, which has no mass and no electric charge but carries energy, and a positron, a particle like an electron but with a positive electric charge.

The proton-neutron pair can then join with another proton to make a nucleus of light helium, which has only one neutron instead of the usual two (2).

Finally, two of the light helium nuclei collide, to form a stable nucleus of helium (3). The two spare protons escape.

Put simply, the Sun can glue together four protons to make a nucleus of helium and generate huge amounts of energy at the same time. The mass of four protons is slightly higher—by 0.5 percent—than the mass of a helium nucleus. The missing mass disappears and energy is created instead. The same or similar fusion reactions between nuclei happen in all normal stars.

Scientists hope they will one day tame nuclear fusion in order to make safe and cheap nuclear power for Earth. In experiments at the European Fusion Laboratory near Oxford, England, and elsewhere, nuclear energy has been released by processes similar to those in the Sun. The challenge for scientists is to use this knowledge to design a safe fusion reactor to generate electricity. Far less nuclear waste would be produced by a reactor that used the same reactions as the Sun than the nuclear power stations we have now.

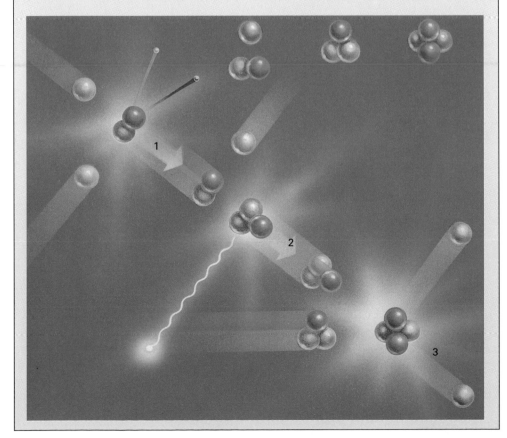

Inside the Sun

It is easy to imagine the Sun as a blazing fire. That is exactly what scientists thought until the 20th century. In 1892 a book stated that the Sun is a mighty furnace of heat and flame. Another theory in the 19th century suggested that meteorites falling on the Sun kept it hot. Both ideas are incorrect. What we now know is that the Sun's furnace is a huge nuclear fire. The energy the Sun brings to planet Earth is made by nuclear reactions deep inside the Sun.

To find out about the Sun's fire, imagine starting at the yellow surface layer, where the temperature is four times hotter than molten iron. At this temperature every known element and substance is a gas, so the entire Sun is a great ball of hot gas.

As we go into the Sun, the temperature and pressure both rise steadily. At every level inside the Sun the pressure of very hot gas pushing outward just balances the crush of gravity pulling in. Toward the core of the sun, the temperature is 25,000 times hotter than at the surface. It is impossible to imagine how hot that really is, but we can write it as between 25,000,000°F and 27,000,000°F.

How long will the Sun last?

The Sun processes about 606 million tons of hydrogen every second, producing more than 4 million tons of helium in the process. By comparing this rate with the mass of the Sun, we can ask: how long will our Sun last?

Obviously, the Sun will not last forever, although it has got an incredibly long life ahead of it. The Sun is now middle-aged. It has taken about 5 billion years to use up half its hydrogen fuel. In the years to come it will slowly get hotter and a little larger. Over the next 5 billion years, its temperature and size will gradually increase as the hydrogen is burned. When all the hydrogen in the central core has gone, the Sun will be three times larger than it is now. On Earth the oceans will boil away. The dying Sun will swallow Earth and turn solid rock into molten lava.

Deep in the Sun, helium atoms will combine to make atoms of carbon and heavier substances. Eventually, the Sun will cool to a ball of nuclear waste, known as a white dwarf star.

LIVES OF THE STARS

A star is a huge ball of hot gas, held together by its own gravity, and heated by nuclear energy. Most stars last for billions of years. They and the planets form from gas clouds in space. Although a great many stars are similar to the Sun, there are also giants as big as the solar system and dwarfs the size of Earth.

Stars come in all ages: newborn, young, middle-aged, and old. New stars are always forming, and old stars are dying. The youngest stars, known as T Tauri stars after the constellation Taurus, are similar to the Sun but much younger. In fact, they are still in the process of forming, and are examples of protostars. They are variable stars whose light output varies because they have not yet settled down to the steady existence of a normal star. Many of the T Tauri stars have swirling disks of material around them and, at the same time, are blowing out powerful "winds" into space.

Material falling into a protostar under the action of gravity has energy that is converted into heat. As a result, the temperature inside the protostar keeps going up. Gradually, it begins to be more like a real star. Then the central region gets so hot that nuclear fusion starts; at this stage the protostar has turned into a normal star. Once nuclear energy starts to be released, the star has a source of energy to keep it going for a very long time. Just how long depends on how big the star is to start with, but a star like our Sun has enough fuel to keep going steadily for about 10 billion years. Stars a lot more massive than the Sun may last only a few million years, however, because they blaze through their nuclear fuel at a very high rate.

△ The Rosette Nebula, a glowing red cloud of hydrogen, surrounds a bright young cluster of stars, NGC 2244.

▽ Formation of a star: (1) Dense clumps of hydrogen and dust form inside a molecular cloud. (2) A protostar forms at the center of a clump. (3) Matter is blown away from the two poles of the star and concentrates around the equator. (4) The new star is surrounded by a disk of dust and gas, which may eventually form a planetary system.

SEE A STELLAR BIRTHPLACE

In the constellation Orion, just below the line of three stars marking the Hunter's belt, you will find the Orion Nebula, a great cloud of glowing hydrogen gas. On a dark night, you may just be able to see the misty patch with the naked eye. Four white-hot stars, called the Trapezium, lie at the heart of the nebula. This quartet of stars floods the surrounding gas with intense ultraviolet radiation, causing it to glow.

Although the Trapezium itself is hard to see, the surrounding gas has a beautiful soft glow, which is quite easy to find with binoculars or a small telescope.

The Trapezium is a place where new stars are being made right now. Astronomers can tell that the Trapezium stars are no more than 1 or 2 million years old. That is very young indeed compared with the Sun, which is already

▷ ▽ The Orion Nebula and its surrounding stars (top), the Trapezium stars (middle), and a jet of matter from a newly forming star in Orion imaged by the Hubble Space Telescope (bottom).

5 billion years old. They are not the only youthful stars in the sky, and there is plenty of evidence that new stars are forming all the time. Our best guess is that several are created every year. But exactly how do they come into being?

The main clues to star birth come from the places where very young stars are found: they are always close to dark interstellar clouds. Photographs of the Orion Nebula, for example, show dark patches as well as the glowing clouds. Lurking behind the Orion Nebula, hidden from direct view, lies a giant molecular cloud. Although it cannot be seen, microwave and infrared signals from the cloud reveal its presence and what is going on inside.

In the giant molecular cloud, dense clumps of gas become the seeds for new stars. Once such a seed has formed, the force of gravity soon pulls more material into it and causes the clump to become denser and denser. It becomes what is called a protostar: not yet a star but on the way to becoming one.

Even before this stage, though, something must make the clumps get started. That can happen when a wave of pressure blasts through the interstellar cloud, perhaps in a collision between galaxies, when giant stars explode, or simply from the intense radiation given off by massive stars. The trick is to get star birth started: once a few young stars are made in a cloud, their action starts the creation of still more stars in the nearby regions. Possibly the Orion Nebula we can see now will disperse into space over the next 10,000 years, but another glowing nebula will appear around stars now in the process of forming.

When stars first form, they are shrouded in dust and gas. Before we can see them in visible light, they must blow away their cocoons. However, the star inside the dusty shell heats up the dust to about 2,000°F. At this temperature, it gives off infrared radiation. Infrared astronomy, which is undertaken from satellites and high mountain observatories, is especially important for studying stars in the process of formation.

Making the planets

By observing stars in the process of formation, we can get a good idea of how our own Sun came into existence. But how did it acquire its family of planets, and do other stars have planetary systems?

As a protostar collapses, it creates a disk of material that surrounds the star. Some of the disk material falls back into the star, pulled in by the star's gravity. The gas and dust that remain in the disk gradually cool down. When the temperature is low enough, it starts to collect together in small clumps, in the same way that raindrops condense out of moist air when the temperature falls. These become planetesimals, the building blocks of planets.

When the solar system formed, some planetesimals broke up as the result of collisions, while others combined to form the planets. In the outer part of the solar system, the large planet cores that formed were able to hold on to some of the gas in the original cloud to form the "gas giants," Jupiter, Saturn, Uranus, and Neptune. They probably developed their own mini-disks of gas and dust, from which their own moons and rings were eventually created.

So far it has not been possible to observe planets around other stars because telescopes simply are not powerful enough. We would not even be able to see the largest planet, Jupiter, if it were orbiting one of the stars nearest the Sun. It is possible to detect "wobbles" in the motion of stars across the sky. These could be caused by the presence of planets, but no conclusive evidence has been found. With the Hubble Space Telescope, and ground-based telescopes under construction, scientists should be able to see if there are planets orbiting some nearby stars.

▷ Formation of the planets, showing three stages that take about 100 million years to occur.
(1) A nebula surrounds the protostar.
(2) Gradually, the nebula cools down and the clumps within it grow, so that 50 million years later the giant gas planets are forming.
(3) Lastly, the rocky matter in the inner solar system forms Mercury, Venus, Earth, and Mars.

Normal stars

All stars are basically like our Sun: they are huge spheres of very hot glowing gas making nuclear energy deep in their interiors. But not all stars are exactly like the Sun. One obvious difference is color. Some stars are reddish or bluish rather than yellow.

Stars also differ in brightness. How bright a star appears in the sky depends on its distance from Earth as well as its actual luminosity. When distance is taken into account, we find that stars cover a range from less than one ten-thousandth the output of the Sun to the equivalent of more than a million Suns. The vast majority of stars, it turns out, are at the dim end of the range. The Sun, though a typical star in many ways, is much more luminous than most. Few of the intrinsically faint stars are visible to the naked eye. The constellations of our skies are dominated by the beacon lights of uncommon, very luminous stars.

Why do stars cover such a range? Everything depends on mass. The amount of material that goes into a particular star determines its color and brightness, and how it changes over time. The minimum amount of mass needed to make a real star is about one-twelfth the Sun's mass. If there were less material than this, the center would never get hot enough for nuclear reactions to be sustained. Between one-hundredth and one-twelfth of the Sun's mass produces an object called a brown dwarf, which briefly generates some energy but cannot become a true star. Because brown dwarfs do not emit any light, they are extremely hard to detect.

At the other end of the scale, astronomers are not certain how heavy the most massive stars are. There seem to be very few stars of more than 60 times the sun's mass and probably none over 100 times more massive than the sun.

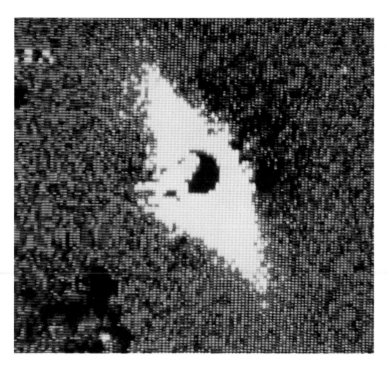

△ Above left, a false-color image obtained from infrared light (heat radiation) shows a disk of dust around the bright star Beta Pictoris. The dust is similar to the material found in the comets of our solar system. This is the best example astronomers have found of a young star with a disk that could become a planetary system. Above right, the structure of the Beta Pictoris gas disk, which is about 60 billion miles in diameter.

THE TEN NEAREST STARS

Name	Distance (light-years)	Magnitude	Constellation
Proxima Centauri	4.2	15.5	Centaurus
Alpha Centauri, A and B	4.4	4.4, 5.7	Centaurus
Barnard's Star	6.0	13.2	Ophiuchus
Wolf 359	7.8	16.7	Leo
BD +36°2147	8.2	10.5	Ursa Major
L 726-8 (A) and UV Ceti (B)	8.4	15.5, 16.0	Cetus
Sirius A and B	8.7	−1.4, 11.2	Canis Major
Ross 154	9.5	13.1	Sagittarius
Ross 248	10.4	14.8	Andromeda
Epsilon Eridani	10.8	6.1	Eridanus

(The letters A and B refer to the two members of a binary star system.)

Giants and dwarfs

The most massive stars are the hottest and brightest. They appear white or bluish. Though large, they produce such enormous amounts of energy that they burn through all their nuclear fuel reserves in just a few million years. By contrast, low-mass stars are dim and reddish in color. They can keep going as ordinary stars for billions of years.

Many of the very brightest stars in the sky, however, are red or orange. They include Aldebaran—the eye of the bull in Taurus, and Antares in Scorpius. How can these cool stars with surfaces that glow rather feebly rival the intense white-hot light of stars like Sirius and Vega? The answer is that they have expanded greatly and are much larger than normal red stars. For that reason, they are described as giants or even supergiants.

Their huge surface areas mean that the giant stars radiate vastly more energy than normal stars like the Sun, even though their surface temperatures are cooler. A red supergiant star—Betelgeuse in Orion, for example—has a diameter several hundred times that of the Sun. In contrast, a normal red star is typically one-tenth the Sun's size. Normal stars are "dwarfs" in contrast to the giants. Giant and dwarf stars occur at different stages in the lives of the stars, and a giant star may eventually turn into a dwarf star when it reaches old age.

The life cycle of a star

An ordinary star, such as the Sun, releases its energy by changing hydrogen into helium in the nuclear furnace at its core. The Sun contains vast amounts of hydrogen, but the supply will not last for ever. The Sun has used up half its hydrogen fuel in the last 5 billion years and can keep going for another 5 billion before supplies of hydrogen run out in the core. What then?

Major changes take place inside a star when it has used up the hydrogen in its central region. Hydrogen away from the center starts to burn in a shell that grows outward. As a result, the star's size increases enormously. At the same time, the surface temperature drops. This process creates red giants and supergiants. It is part of the sequence of change called "stellar evolution," which all stars undergo.

Ultimately, all stars reach old age and die, but a star's particular mass is crucial to how long it lives. Massive stars dash through their life cycles, ending it all in a spectacular explosion. In contrast, more modest stars, including the Sun, end up collapsing into extremely dense white dwarfs. Then they just fade away.

In the process of changing from a red giant to a white dwarf, a star may blow off its outer layers as a hollow shell of gas, leaving its core naked. The shell of gas glows brightly in the intense radiation from the star that is left, which may be as hot as 180,000°F at its surface. Such glowing gas bubbles were termed planetary nebulae when they were first seen, because they usually appear to be circular, like the disk of a planet in a small telescope. In reality, of course, they have nothing to do with planets!

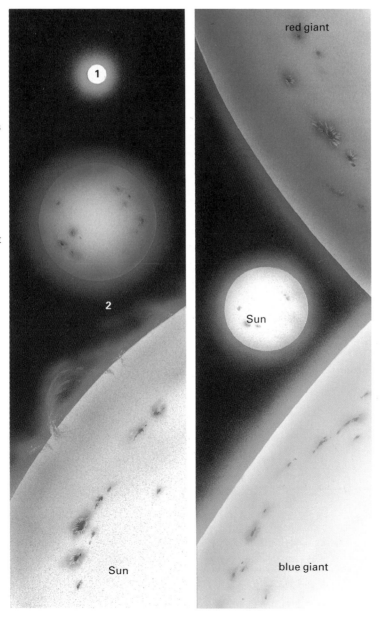

▷ The Sun compared in size with red and white dwarf stars (left) and giant stars (right). Astronomers find star diameters by comparing the temperature of a star's surface with its brightness: the larger the diameter of a star, the more energy it radiates at a given temperature. A white dwarf star is about the same size as Earth, but contains a million times as much matter. At their greatest size, red giant stars are larger than the diameter of Earth's orbit around the Sun.

1. white dwarf
2. red dwarf

red giant

Sun

Sun

blue giant

▷ The life cycles of two different stars. The top half of the diagram shows the stages from birth to death of a star massive enough to explode as a supernova and end up either as a neutron star or as a black hole (see pages 120–123). The lower half shows the stages in the life of a less massive star, like our Sun. Material blown off stars as they evolve is recycled into nebulae from which future stars form. The broad blue arrows represent this process.

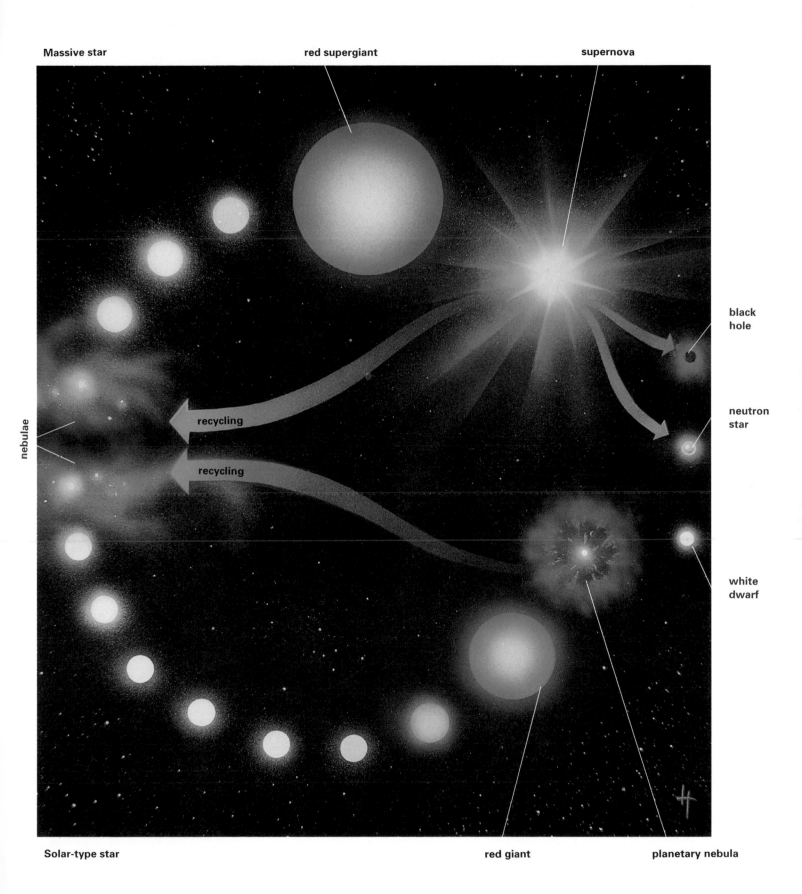

Massive star

red supergiant

supernova

black hole

neutron star

nebulae

recycling

recycling

white dwarf

Solar-type star

red giant

planetary nebula

THE MAIN SEQUENCE

Small stars are dim and red. Large stars are very luminous and blue-white. Our yellow Sun comes in between. If you choose a sample of stars at random from the sky and plot a point for each one on a graph with color along one axis and energy output along the other, most of the points will fall along a diagonal line. This line was called the main sequence by the first astronomers to draw such a diagram. Points corresponding to giants (**2**) and supergiants (**1**) lie above the main sequence and white dwarfs (**3**) below it. We now know that a main-sequence star is a normal star, like the Sun, which is burning hydrogen in its core.

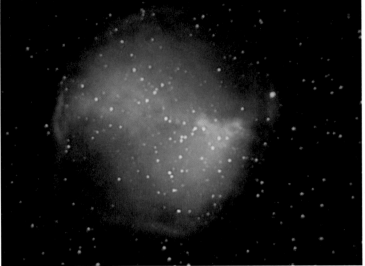

◁ A planetary nebula, M27. This star is nearing the end of its life and has thrown off its outer layers. The star at the center of the nebula has a surface temperature of 180,000°F. The expanding shell will gradually merge with interstellar matter.

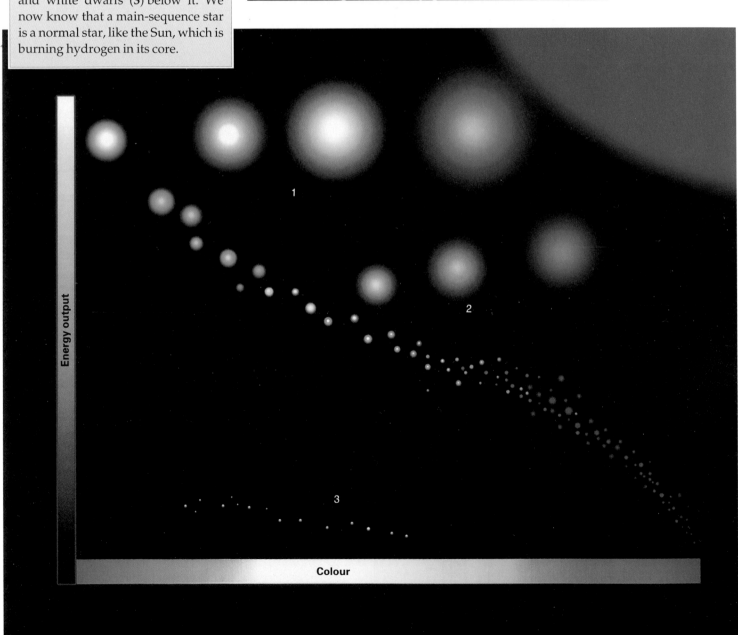

Energy output

Colour

Star clusters

Nearly all stars seem to be born in groups rather than as individuals. It is not surprising, then, that star clusters are common. Astronomers like to study clusters because they know all the stars within a cluster were formed at roughly the same time and are about the same distance away from Earth. Any apparent differences in brightness between the stars are real differences. In whatever way the stars have changed over time, they started off the process together. Star clusters are particularly useful for seeing the variations in properties among stars with different masses because all are of the same age and at the same distance, with mass being the only difference between them.

As well as being interesting for scientific study, star clusters are also among the most beautiful subjects for astronomical photographs and for viewing by amateur astronomers. There are two types: open clusters and globular clusters. The names describe their appearance in the sky. In an open cluster, individual stars can be seen separately, scattered over a patch of sky. By contrast, globular clusters are spheres of stars so densely packed that individual members cannot be picked out in the central region.

Open clusters

Perhaps the most famous of all open clusters is the Pleiades or "Seven Sisters," which lies in the constellation Taurus. Despite its name, most people can pick out only six stars without the help of a telescope. The total number of stars in the cluster is somewhere between 300 and 500, all within a region 30 light-years across and 400 light-years away. The cluster's age is only about 50 million years—young by astronomical standards—and it contains very massive luminous stars that have not yet turned into giants. The Pleiades is typical of open clusters, which have between a few hundred and a few thousand member stars.

Hardly any open clusters are known to be older than 100 million years, and there are more young ones than old ones. It is unlikely that the rate at which they form has changed. Rather, the stars in older clusters gradually get separated from each other until they join the general population of noncluster stars, thousands of which we can see in the night sky. Though gravity holds the open clusters together to some extent, they are quite fragile and the pull from another object, such as a large interstellar cloud, might easily tear a cluster apart.

Some star groups are held together so loosely that they are called stellar associations rather than clusters. Because they do not last very long, the ones we see are made up of very young stars, close to the interstellar clouds where they formed. Associations have between 10 and 100 members, which may be scattered over a region several hundred light-years across.

Star-forming clouds are concentrated in the disk of our Milky Way Galaxy, and this is where open clusters are found. Against the vast star clouds of the Milky Way, and with so much dust in interstellar space, the 1,200 open clusters we know of must be just a small fraction of the real total for the galaxy. Perhaps there are as many as 100,000 in all.

◁ The Pleiades (NGC 1432) is a young cluster of stars, clearly visible to the naked eye in the constellation Taurus. There are about 300 stars in this cluster and it is 400 light-years away. The stars are inside a nebula of cold gas and dust, which appears as blue wisps in this photograph.

▷ The double cluster in Hercules, a pair of open clusters. Notice how spread out the stars appear in both clusters.

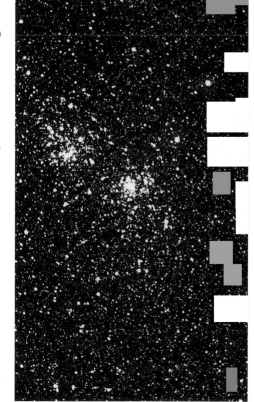

Globular clusters

In marked contrast to open clusters, globular clusters are densely packed spheres of hundreds of thousands or even millions of stars. The stars are so crowded that, if our Sun were in a globular cluster, more than a million individual stars could be seen in the night sky with the naked eye. Typically, cluster sizes lie between 20 and 400 light-years.

In the cramped cores of these clusters, stars pass so close to each other that their mutual gravitational pull ties them together to make close binary stars. Some even merge together completely; close encounters can shred away the outer layers of a star, revealing the central nuclear core to direct view. Double stars are 100 times more common in globular clusters than elsewhere. Some of these binaries are sources of X-rays.

About 200 globular clusters are known around our Galaxy, distributed throughout a large spherical halo all around the Galaxy. All are very old and formed more or less at the same time as the Galaxy itself, about 10 to 15 billion years ago. It is likely that they were created when parts of the cloud from which the Galaxy formed fragmented into smaller pieces. The clusters have not dispersed because their stars are so densely packed and the strong gravitational pull they all have on each other keeps the clusters tightly together.

Globular clusters are also seen around other galaxies of all kinds, not just our own. The brightest globular cluster, easily visible to the naked eye, is Omega Centauri in the southern constellation Centaurus. It is 16,500 light-years away from the sun and has a diameter of 620 light-years, making it the largest known. The brightest globular cluster in the Northern Hemisphere is M13 in Hercules, which is just discernible to the naked eye.

▷ The giant globular cluster Omega Centauri (NGC 5139) is 16,500 light-years from the Sun. It is 620 light-years across, making it the largest globular cluster known in our Galaxy. It is billions of years old and contains hundreds of thousands of stars. From the Southern Hemisphere, Omega Centauri is easily visible to the naked eye.

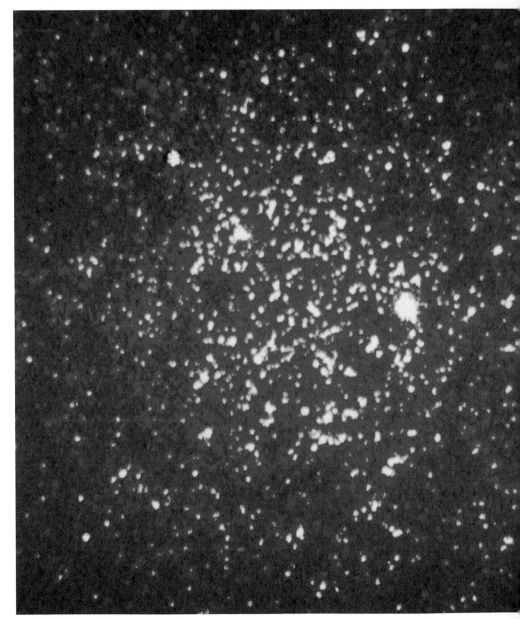

▷ With an ultraviolet telescope on a spacecraft, the central stars of the cluster Omega Centauri are seen. This view cannot be obtained by telescopes on Earth's surface because the atmosphere blurs the star images too much.

FIND A STAR CLUSTER

The diagram below shows where to find the Pleiades cluster, an unmistakable little group of stars. Choose a dark night between about November and March. In addition to the Pleiades, there are several well-known open clusters that are easy to find. The Hyades, also in Taurus, is the nearest to us at a distance of 150 light-years. Its stars are loosely grouped around the star Aldebaran. Praesepe (Latin for "the manger"), also called the Beehive in English, lies at the center of Cancer; there is also a unique double cluster in Perseus.

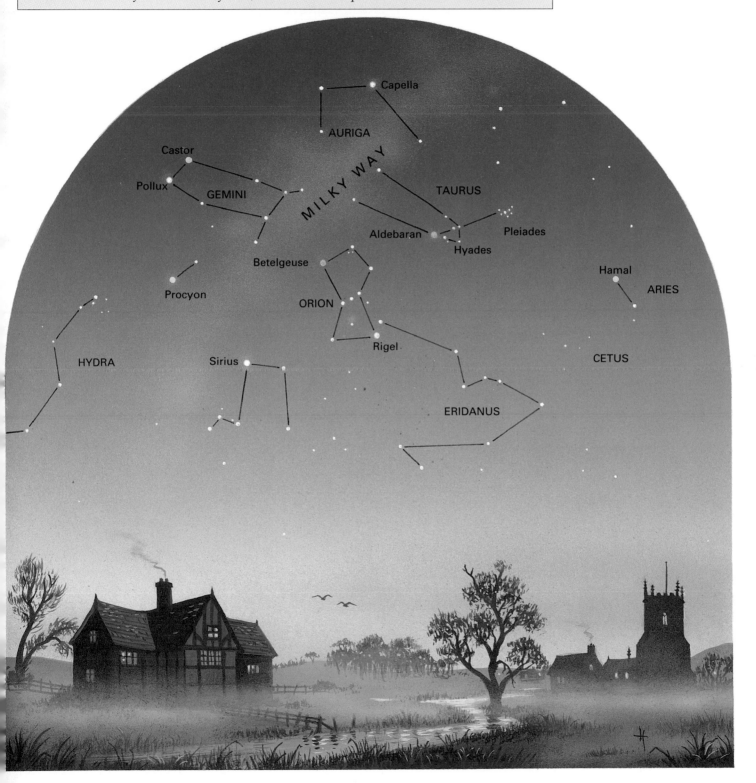

VARIABLE STARS AND BINARY STARS

Although the Sun shines with the same brightness day after day, elsewhere in our Galaxy there are stars that vary in their brightness. The ones that vary in a predictable way are used to work out distances in our Galaxy and beyond.

In 1596 a Dutch amateur skywatcher named David Fabricius (1564–1617) noticed a fairly bright star in the constellation Cetus (the Whale) gradually get dimmer over several weeks until it finally vanished from sight. He had made the first recorded observation of a variable star.

The star became known as Mira—the "wonderful star." Over a period of around 332 days, Mira varies in brightness between about 2nd magnitude (as bright as the polestar) and 10th magnitude, when it is far too dim to be seen with the naked eye. Today many thousands of variable stars are known, though most do not change as dramatically as Mira.

Why some stars vary

There are many different reasons why stars vary in brightness. They can vary by many magnitudes, or by so little that only sensitive instruments can detect the change. Some stars vary in a regular or nearly regular way. Others fade unpredictably or have sudden outbursts. The changes may take place over a cycle of several years or happen in a matter of seconds.

To understand why a star is variable, it is important first to track exactly how it varies. A graph showing the way in which the magnitude of a variable star changes over time is called a light curve. To plot a light curve, brightness measurements have to be made regularly. Professional astronomers use an instrument called a photometer to measure the magnitudes of stars accurately, but many observations of variable stars are made by amateurs. With the help of a specially prepared map and some practice, it is not too difficult to judge the magnitude of a variable star with the naked eye when it is compared with nonvariable stars nearby in the sky.

Graphs of variable stars show that some types vary in a regular way, which repeats again and again over a period of time, and others are rather unpredictable. The regular variable stars include pulsating stars and binary stars.

The light output of some individual stars varies because they pulsate or throw off clouds of matter. But there is another group of variables that are double, or binary, stars. In binary stars, several different things can happen to make us see a change in brightness. The two stars may be aligned from our point of view so that they cross in front of each other as they orbit. Systems like this are called eclipsing binaries. The most famous example is Algol in the constellation Perseus. In a close binary pair, material can stream from one star to the other, with dramatic effect.

Pulsating variable stars

Some of the most regular variable stars pulse in and out, vibrating at a particular frequency in much the same way as a

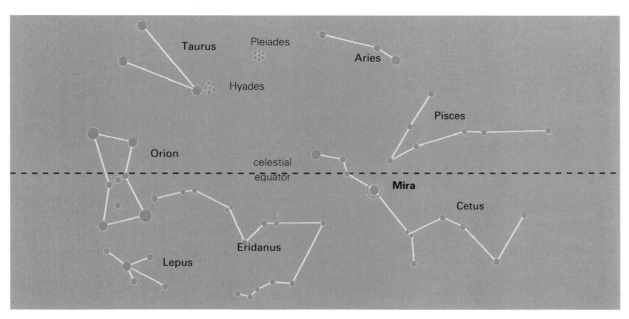

◁ The position of the variable star Mira in the constellation Cetus, the Whale. Sometimes it is the brightest star in its constellation. At other times it is too faint to see with the naked eye.

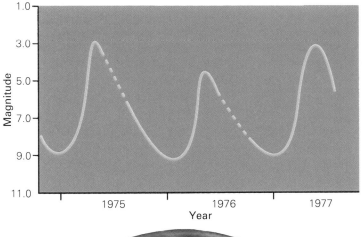

◁ This illustration shows how the brightness of the variable star Mira changes over time. The line is dotted during periods when Mira is not visible in the night sky.

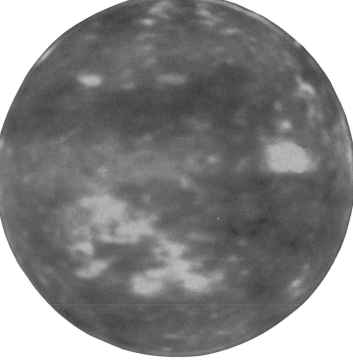

◁ The turbulent surface of a flare star might look like this from close by. In a stellar flare, energy is released suddenly in the space of one minute or less, and the temperature goes up by several thousand degrees.

magnitudes, and then it gradually returns to normal. The explanation seems to be that the supergiant star blows off clouds of carbon that condense into grains, like soot. If one of these thick black clouds comes between us and the star, its light is blotted out until the cloud has dispersed into space. Stars of this kind produce the thick dust that is so important in regions of star formation.

Flare stars

On the Sun, magnetic effects cause sunspots and solar flares, though these do not change the Sun's brightness significantly. In some red dwarf stars, however, flares generated in a similar way reach huge proportions, resulting in a sudden increase in light output by a magnitude or more. The nearest star to the Sun, Proxima Centauri, is one such flare star. The outbursts cannot be predicted in advance and they last only a few minutes.

string on a musical instrument. Cepheids are the best-known pulsators, named after Delta Cephei, which is a typical example. These are supergiant stars, between three and ten times more massive than the sun and hundreds or thousands of times more luminous. As a Cepheid vibrates over a period of days, both its surface area and surface temperature change, affecting its overall brightness.

Mira, the first recorded variable star, and other similar stars also owe their variability to pulsations. They are cool red giants in a late stage of their lives and may be about to throw off their outer layers altogether as a shell to create a planetary nebula. Most red supergiants, like Betelgeuse in Orion, vary to some extent

in magnitude. By using special observation techniques, astronomers have revealed large dark spots on its surface.

RR Lyrae stars are another important group of pulsators. They are old stars of about the same mass as the Sun. Many are found in globular clusters. They typically vary by about one magnitude in something under a day. Like Cepheids, their properties can be used to deduce astronomical distances.

Irregular variable stars

R Coronae Borealis, and stars like it, behave in a completely unpredictable way. Normally, this star is just about visible to the naked eye. Every few years its brightness plunges by around eight

△ The orbit of this binary star, called Krüger 60, in the constellation Cepheus is face on to us. The two stars take 44 years to travel around each other. These three photographs were taken in 1908, 1915, and 1920.

Binary stars

About half of all stars in our Galaxy belong to double-star systems, so binaries (two stars orbiting together) are very common. Belonging to a binary can greatly affect how a star evolves and changes, particularly if the two companions are close. Matter streaming from one star onto another creates dramatic outbursts, such as nova and supernova explosions.

Binaries are held together by their mutual gravitational pull. The two stars swing in elliptical orbits around a point in space between them called their center of gravity. This would be the balancing point if you could imagine the stars each on either end of a seesaw. The farther apart the stars are, the longer it takes them to complete their orbits.

Most binaries are too close for the individual stars to be visible, even with powerful telescopes. If the separation is great enough, the orbital period is many years, perhaps a century or more. Double stars that can be seen are called visual binaries.

Discovering binary stars

Binary stars are most often detected either from the unusual motion of the brighter star of the two, or from the combined spectrum. If a single star seems to wobble across the sky in a regular way, that means it has an unseen companion. It is said to be an astrometric binary, discovered through measurements of its position.

Spectroscopic binaries are detected by changes and peculiar features in their spectra. The spectrum of an ordinary star, such as the Sun, is like a continuous rainbow crossed by numerous narrow gaps, called absorption lines. The precise colors at which the lines occur change if the star is moving toward us or away from us, a phenomenon called the Doppler effect.

As the stars in a binary system go around their orbits, they alternately move toward us and away from us. As a result, the lines in their spectra move about over a range of colors. Such moving spectrum lines signify a binary star. If the two members are similar in brightness, two sets of lines can be seen. If one star is much brighter than the other, its light will dominate, but the regular shift of its spectrum lines gives away its true binary nature.

Measuring the speeds of binary star members, and applying the laws of gravity, is an important method of finding star masses. Studying binaries is the only

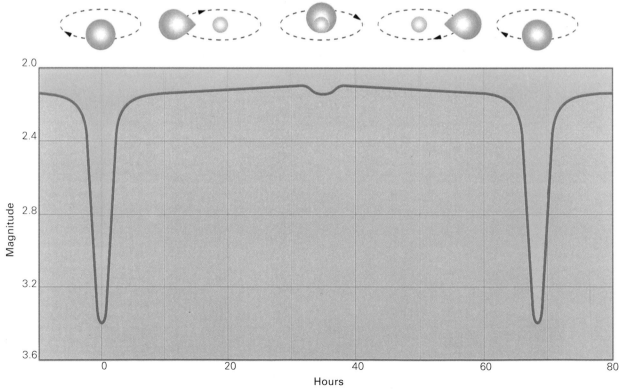

◁ Algol, the "winking" star, is the second brightest in Perseus. Every 69 hours it dips in brightness by a magnitude. Decline takes four hours and the minimum level lasts just 20 minutes before the five-hour rise to normal brightness starts. In reality, Algol is two stars, a hot blue-white star and a dimmer yellow star. The hot star "winks" when the dimmer star crosses in front of its companion. There is a tiny dip in brightness when the yellow star disappears from view, but it is too small to see with the naked eye.

direct way to find the masses of stars. Even so, it is not easy to get precise answers in every case.

Close binary stars

In a close binary the gravitational pull of the pair of stars tends to stretch each star into a pearlike shape. If the pull is strong enough, there comes a critical point where matter can stream off one star and start falling onto the other. Around the two stars, there is a region shaped like a three-dimensional figure eight, the surface of which acts as a critical boundary. The two pear-shaped parts, one around each star, are called Roche lobes. If one of the stars grows large enough to fill its Roche lobe, material streams onto the companion at the point where the lobes touch. Often the material does not land directly, but first spirals around, forming what is known as an accretion disk. If both stars expand to fill their Roche lobes, a contact binary forms. The material of the two stars gets mixed together in a large ball around the two stellar cores. Since all stars eventually swell up into giants, and many stars are double, interacting binaries are not uncommon.

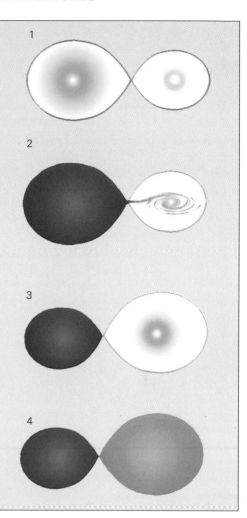

THE LIFE OF A CLOSE BINARY STAR

The famous eclipsing binary, Algol, is a good example of a pair of stars that have swapped material. If the two stars had been single, like our Sun, their evolution would have been completely different.

1 Algol started as a pair of stars in which the larger was 3 times the mass of the Sun, and the smaller 1.5 solar masses.

2 The larger star used up its reserves of hydrogen far more quickly than its companion and gradually expanded to become a red giant, filling its Roche lobe. The red giant started shedding material onto its partner.

3 Now the two stars have changed roles. What started out as the larger star is less massive than the Sun, while the mass of the other has grown to 3.7 solar masses. A very small amount of material is still flowing between them.

4 In the future the other star will also expand until the two are in contact and shared material will envelope them both.

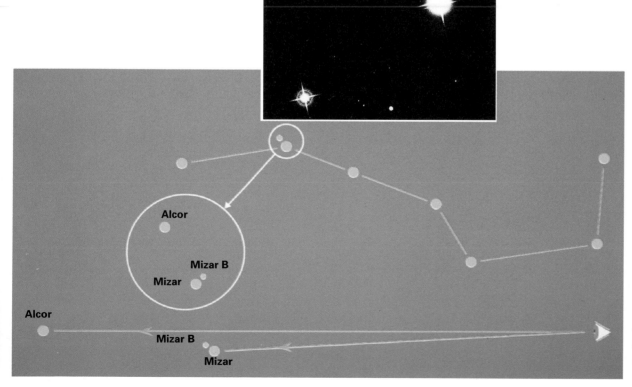

◁ The stars Mizar and Alcor in Ursa Major (the Great Bear) look close to each other in the sky. But in reality, they are 10 light-years apart. However, Mizar does have a real companion, Mizar B. In the photograph (inset) two sets of "spikes" are just visible. The spikes are caused by mirror supports in the telescope, but here they help to show the presence of another star. The spectra of Mizar and Mizar B reveal that both are spectroscopic binaries, so the whole star system has four members altogether.

A star spills over

Nova outbursts are one of the spectacular results of mass transfer in binary stars. One star swells to fill its Roche lobe (which you can picture as the outer layers ballooning to the point where the other star's gravity grabs hold of it). The partner star is a tiny white dwarf. A nova brightens suddenly by about 10 magnitudes.

What has happened is an enormous energy release in a very short time as a massive nuclear explosion is triggered on the surface of the white dwarf. As material from the swelling star streams down to the dwarf, the pressure increases in the cascading material and the temperature under the new layer rises to around 1,800,000°F.

Some novae have been seen to have repeated outbursts after tens or hundreds of years. Others have only been seen to explode once, but such explosions may recur after thousands of years. Another type, called dwarf novae, have less dramatic outbursts that repeat after days or months.

△ The star at the center of these two photographs lies in the constellation Cygnus and exploded as a nova in 1975. The photos show the nova at its peak brightness (top) and some months later (arrowed), after it had faded again.

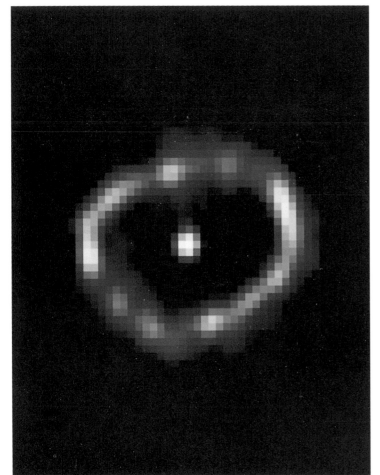

◁ This star in the constellation Cygnus erupted as a nova in January 1994. The bright ring is the edge of a bubble of gas blasted off by the nova. It was observed by the Hubble Space Telescope in January 1994.

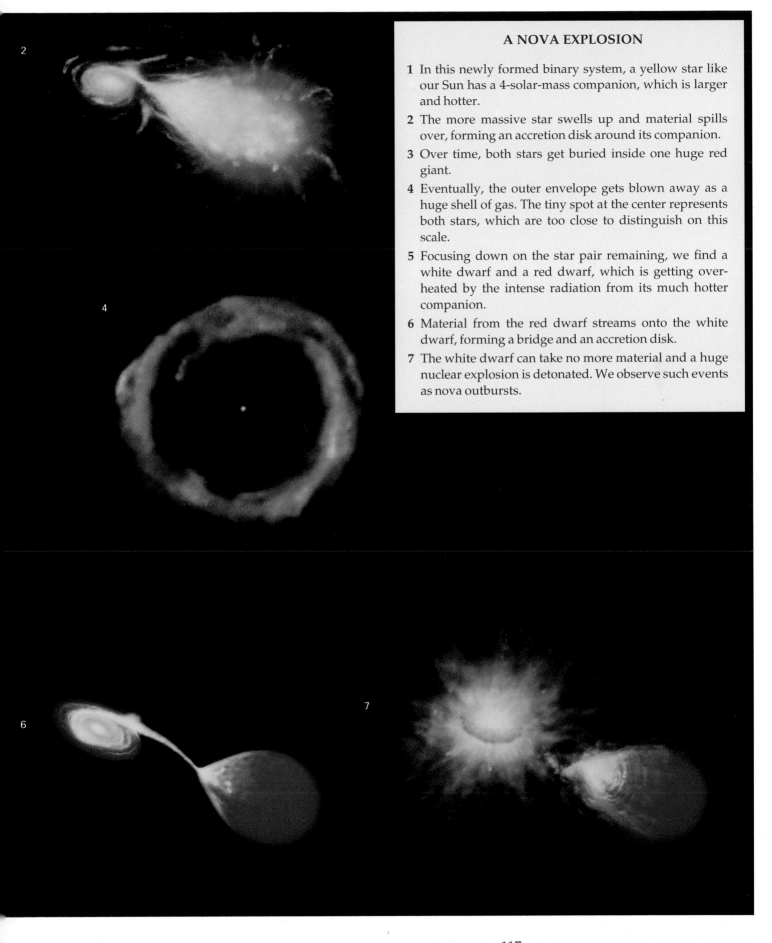

A NOVA EXPLOSION

1 In this newly formed binary system, a yellow star like our Sun has a 4-solar-mass companion, which is larger and hotter.

2 The more massive star swells up and material spills over, forming an accretion disk around its companion.

3 Over time, both stars get buried inside one huge red giant.

4 Eventually, the outer envelope gets blown away as a huge shell of gas. The tiny spot at the center represents both stars, which are too close to distinguish on this scale.

5 Focusing down on the star pair remaining, we find a white dwarf and a red dwarf, which is getting overheated by the intense radiation from its much hotter companion.

6 Material from the red dwarf streams onto the white dwarf, forming a bridge and an accretion disk.

7 The white dwarf can take no more material and a huge nuclear explosion is detonated. We observe such events as nova outbursts.

EXOTIC STARS

When a star has used up all its hydrogen fuel, gravity takes over and the star shrinks. At the end of its life cycle, the exhausted star may fade quietly as a white dwarf or dramatically as an exploding star. Explosions of giant stars may create small black holes in space.

When a star has used up all its nuclear fuel and is no longer releasing new energy at its center, it starts to collapse in on itself. The pull of gravity inward is not balanced anymore by the outward push of hot gas. What happens next depends on the mass of the collapsing material.

If the mass is less than 1.4 times the mass of our Sun, the star stabilizes as a white dwarf. The catastrophic infall is halted by a basic natural property of electrons. There comes a point where they can push back, even though there is no source of heat. However, this does not happen until the electrons and atomic nuclei are crammed together to create an extremely dense kind of matter.

A white dwarf with the Sun's mass is only about the size of Earth. Just a cupful of white dwarf material would weigh a hundred tons on Earth. A curious fact about white dwarfs is that the more mass they have, the smaller they are.

It is very difficult to imagine what a white dwarf is like inside. It is like a single gigantic crystal that is gradually cooling off. It will get progressively dimmer and redder. In fact, though astronomers refer to white dwarfs as a group, only the hottest examples, with surfaces at about 20,000°F, are really white. Ultimately, every white dwarf will turn into a dark ball of radioactive ash, the totally dead remains of a star.

White dwarfs are so small that even the hotter ones do not give out much light and they are not very easy to find. Even so, hundreds are now known and astronomers estimate that as many as one-tenth of all the stars in the Galaxy could be white dwarfs. Sirius, the brightest star in the sky, has a white dwarf companion called Sirius B.

Neutron stars

From our knowledge of physics, it is possible to calculate that a collapsing star of more than 1.4 solar masses does not stop at being a white dwarf. Gravity is so strong that the electrons get forced inside the atomic nuclei. The effect is to turn protons into neutrons (see pages 20–21), which are crammed together with no space between. Neutron stars are even denser than white dwarfs but, like electrons, the neutrons themselves can stop the collapse as long as there is no more than about 3 solar masses of material. A typical neutron star is only 5 to 10 miles across and a cubic inch of it would weigh a billion tons.

In addition to their enormous density, two other special features of neutron stars make it possible to detect their presence even though they are so small: their rapid spin and their strong magnetic field. All stars rotate, but as a star collapses, its rate of spin goes up dramatically, in just the same way as a skater spins faster by drawing in his or her arms. A neutron star can rotate many times each second. Combined with this extremely rapid rotation, neutron stars have magnetic fields a million million times stronger than Earth's.

Pulsars

The first pulsars were discovered in 1968, when radio astronomers detected very regular blips coming from four locations in our Galaxy. They were amazed that something natural could emit radio pulses in such a regular and rapid manner. For a short time, the astronomers considered the possibility that intelligent beings elsewhere in the Galaxy were responsible. But soon there was a natural explanation.

In the intense magnetic field of a neutron star, spiraling electrons generate

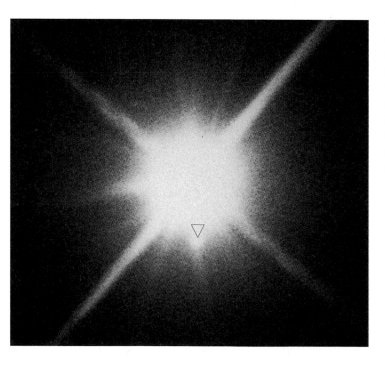

◁ The faint white dwarf companion of Sirius, the brightest star in the sky, is just visible in this photograph. It is called Sirius B. The pair take 50 years to complete one orbit around each other.

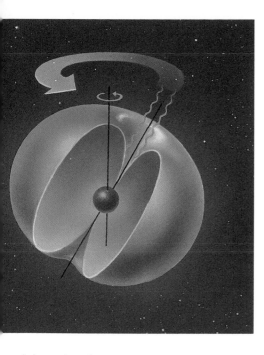

radio waves that are beamed out like a searchlight. The star spins rapidly, swinging the radio beam across our line of sight, like a lighthouse. Some pulsars emit light, X-rays, and gamma rays as well radio waves. The slowest pulsars have pulses about four seconds apart but the most rapid repeat in a matter of milliseconds. The rotation of these neutron stars has been speeded up by some means, probably because they are in binary systems.

X-ray binary stars

At least 100 bright sources of X-rays are found in our galaxy. X-rays are so energetic that something exceptional has to take place to make them. Astronomers think that matter falling onto the surface of a tiny neutron star would emit X-rays.

Astronomers accept that these X-ray sources are binary stars in which one member is a very small but high-mass star —in many cases, a neutron star, though some may contain a white dwarf or even a black hole. The companion can either be a massive star of 10 or 20 solar masses or a star of less than 2 solar masses. Anything in between seems to be exceedingly rare. A complicated history of evolution and mass exchange between the star pairs leads to these situations. The final result depends on the starting masses and the initial separation between them.

In the low-mass binaries, a disk of gas forms around the neutron star. In the high-mass case, material streams directly onto the neutron star, funneled by its magnetic field. They are often X-ray pulsars as well.

△ In a pulsar, the magnetic field of the neutron star is tilted at an angle to its rotation axis. As the star spins, the radio waves beaming along the magnetic axis sweep around like a search-light. We detect radio pulses when the beam points at Earth.

▷ When one star in a binary pair is a neutron star, enormous amounts of energy are given off in the form of X-rays as matter falls into the neutron star from its companion. There are two kinds of X-ray binaries: the neutron star's companion can either be a white dwarf or a massive blue giant.This picture shows both kinds to scale. The size of the Sun is shown for comparison and the white dwarf type is also shown enlarged.

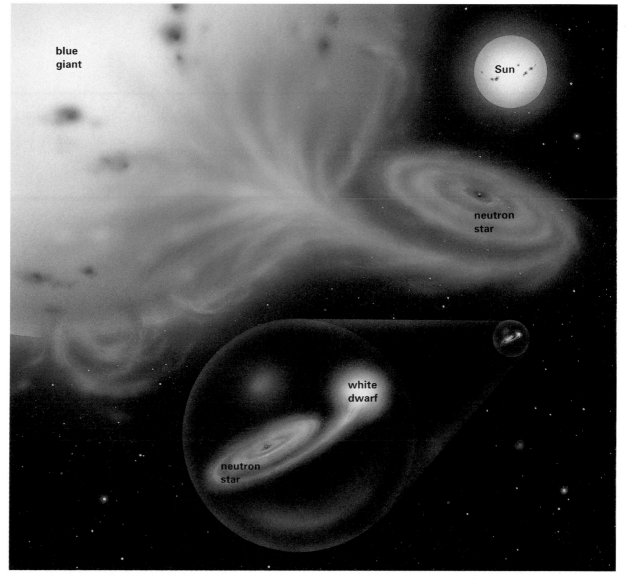

blue giant

Sun

neutron star

white dwarf

neutron star

Black holes

According to theoretical calculations, the mass of a neutron star cannot be greater than about 3 solar masses. So what happens if a more massive star collapses? The favored idea is that it probably becomes a black hole.

Black holes have strange properties that give them a special fascination. In a region around the collapsed mass, the gravitational field is so strong that not even light can escape. The boundary of this region is called the event horizon because an outside observer cannot see anything happening beyond it. Inside, there is nothing—except perhaps something in physics not yet fully understood—to stop the material from collapsing to an infinitely small point. There is nothing actually there at the event horizon. A doomed astronaut falling through would not notice anything in particular, but it acts as a kind of one-way valve. Anything can fall through it, but nothing can get out again. For a black hole of 3 solar masses, the event horizon has a radius of 5.5 miles.

Do black holes actually exist? Almost certainly they do. In a number of binary systems, where it is possible to weigh the stars by studying their motion, very compact stars have been found that appear too massive even to be neutron stars. In one X-ray binary, called A0620-00, the mass of the compact star has been measured very accurately by combining several different kinds of observations. It turns out to be 16 solar masses, far too high to be a neutron star. Another X-ray binary, V404 Cygni, has a black hole of at least 6.3 solar masses.

As well as black holes with stellar masses, supermassive black holes almost certainly exist at the centers of galaxies. Only material falling into a black hole could account for the stupendous energy output of galaxies with very active nuclei.

Supernovae

Stars of less than 1.4 solar masses fade away quietly when they die. But what happens to more massive stars? And how are neutron stars and stellar black holes created?

The explosive end of a massive star is truly spectacular. It is the most powerful natural event ever seen in stars. In a few moments, more energy is released than our Sun will radiate over 10 billion years. The light output of the single dying star is equivalent to that of a whole galaxy, and what we see as light is only a small fraction of the total energy. Fragments of the original star are blown outward at speeds as high as 12,000 miles per second.

These stupendous stellar explosions are known as supernovae. Supernovae are fairly rare. About 20 or 30 are found in other galaxies each year, mostly as a

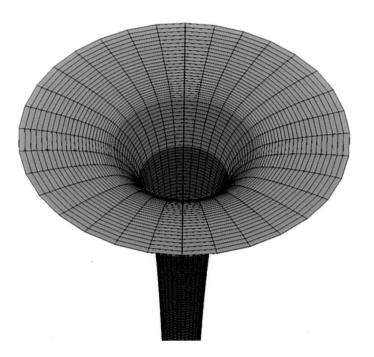

◁ This funnel-shaped well represents the gravitational influence of a black hole. Being pulled by a black hole is like being drawn into the center of a whirlpool. Anything in the outer region (lightest color) can stay at the same distance from the black hole. Anything in the middle region can escape from the black hole as long as it moves fast enough in the right direction. Not even light can escape from the innermost region (darkest color).

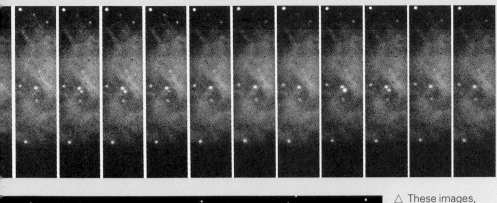

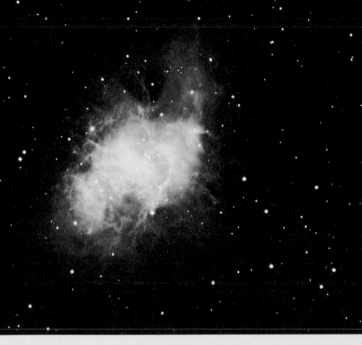

△ These images, taken at very high speed, capture a flash from the pulsar in the Crab Nebula. The pulsar is at the center of each panel.

◁ The Crab Nebula photographed with the five-meter Hale Telescope at Mount Palomar, California.

result of systematic hunting. In any one galaxy, there might be between one and four in a century. In our own Galaxy, however, no supernova has been noted since 1604. Probably there have been supernovae, but we have been unable to see them because of the large amount of dust between the stars in the Milky Way. Radio astronomers discovered a ring of gas left over from a supernova in the constellation Cassiopeia and have been able to work out its explosion date to around 1658. No one recorded an exceptionally bright star at the time, though a modest star that can no longer be seen was recorded in the same area on a star map in 1680.

Supernova star death

To understand what triggers a supernova, we have to look at the last stages in the evolution of a massive star. When the hydrogen in the central core has all been used up to make helium, new nuclear processes start, turning helium into carbon. Farther out from the center, in a shell, hydrogen can still be fusing into helium. When the helium is used up, the carbon becomes the fuel. A whole series of different nuclear reactions take place in layers around the core until the star has a structure somewhat like that of an onion.

The final stage comes when the star develops a core of iron and nickel with shells around it burning silicon, neon,

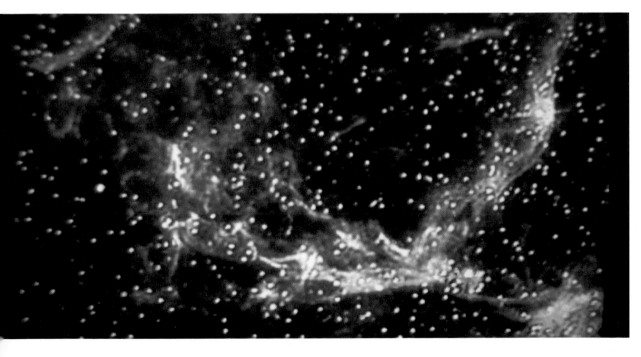

◁ The Veil Nebula in the constellation Cygnus is part of a huge shell of gas blown off in a supernova about 30,000 years ago.

oxygen, carbon, and helium as nuclear fuels. It builds a white dwarf at its center until the maximum mass allowed of 1.4 solar masses is exceeded. Then catastrophic collapse of the core follows rapidly. In less than a second, the core goes from being about the size of Earth to less than 60 miles across. Its density becomes as high as the nucleus of an atom (about 1,000 million million million times denser than water). Material merges like a single gigantic atomic nucleus and a neutron star is created.

At the point where the neutrons in the inner part of the core are able to prevent further collapse, the process comes to an abrupt halt. Shock waves rebound through material still falling in and energy is injected into the star by vast numbers of particles called neutrinos. The effect is to blow off the outer layers of the star, leaving the neutron star core behind, now exposed to view. Astronomers think that most, if not all, neutron stars are formed in supernovae. In some circumstances, the core may be massive enough to become a black hole rather than a neutron star.

We have a good picture of how massive stars end their lives as supernovae—but this is not the only way a supernova is triggered. Only about one-quarter of all supernovae are of this kind. They are distinguished by their spectra and the particular pattern of how they brighten and fade. How the other supernovae work is less certain. The favorite theory is that they start as white dwarfs in binary systems. Material flows from the companion to the white dwarf until the 1.4-solar-mass limit is passed. In the supernova explosion that follows, it is thought that the whole star is probably destroyed.

A supernova stays at its peak brightness for only about a month before declining continuously. During this time, the light is powered by the radioactive decay of material created in the explosion. Long after a supernova has exploded, it is still possible to see the shell of material it blew off, which gradually expands into space. These nebulae are called supernova remnants. In the constellation Taurus, the Crab Nebula is the remains of a supernova observed in 1054. A large, thin ring of material in Cygnus, called the Veil Nebula or the Cygnus Loop, is left over from a

SUPERNOVA 1987A

When SN 1987A was discovered on February 24, 1987, it created great excitement among astronomers because it was the brightest supernova to be seen since 1604. Though in our neighboring galaxy, the Large Magellanic Cloud, it reached magnitude 2.9 at its peak, easily visible to the naked eye from countries in the Southern Hemisphere. This was the first time modern instruments could be used to study the progress of a supernova. By studying photographs taken beforehand, it was even possible to identify the star that exploded. It turned out to be a blue supergiant star of about 17 solar masses and thought to be about 20 million years old.

The actual explosion seems to have been about a day before discovery. It was found on an earlier photograph and researchers looking for neutrinos from space detected an exceptional number on February 23.

Neutrinos are elementary particles with hardly any mass. They are very difficult to detect but extremely important because of the huge amount of energy they carry when produced in vast numbers in nuclear reactions. Finding the neutrinos confirmed that our theory explaining the supernova is basically correct. However, no pulsar or neutron star has been discovered where this supernova exploded.

◁ ▽ Supernova 1987A blazes in the Large Magellanic Cloud (left) where there is only a dim star of 12th magnitude on a photograph taken previously (below).

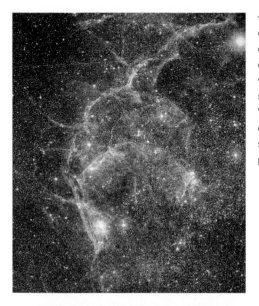

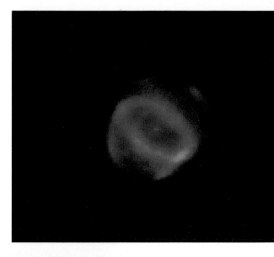

◁ This tangled mass of gas is the remnant of a supernova that exploded about 10,000 years ago. It lies in the southern constellation Vela (the Sail) and contains one of the strongest radio pulsars.

▷ An old star gently blows off most of its outer layers to make a planetary nebula. This is one way in which chemical elements made in stars are recycled into interstellar gas clouds. In the future, this gas may be incorporated in a new star.

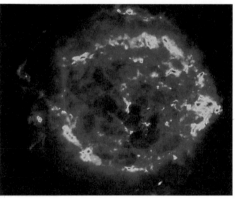

◁ Cassiopeia A is the strongest radio source in the sky after the Sun. It is the remnant of a supernova that exploded in around 1667. No one saw it because it was hidden behind thick dust clouds. This is a color-coded radio map.

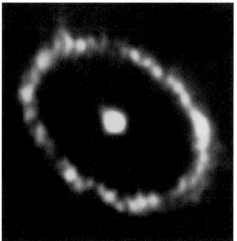

◁ A clumpy ring of glowing gas blasted out into interstellar space by Supernova 1987A when it exploded. In this picture, taken four years after the explosion, it is already 1.37 light-years across. It will continue to expand and will eventually mix into the interstellar gas.

supernova about 30,000 years ago. Supernova remnants are some of the strongest radio sources in the sky.

Origin of the elements

Nearly 100 different chemical elements contribute to our familiar world: rocky Earth with its oceans and atmosphere, plant and animal life.

In the universe, some of these elements are much more plentiful than others. The cosmic elements combine to make innumerable different substances. But where did the basic chemical building blocks of the universe come from? Astronomers are now able to piece together a picture of how different elements are distributed throughout the universe and a history of how they were created. (See also pages 20–21.)

The simplest of all the elements is hydrogen. An atom of hydrogen has a single proton as its nucleus and one electron to make up the atom. Different elements contain different numbers of protons in their nuclei and all elements except hydrogen also have a number of neutrons.

In nuclear reactions, atomic nuclei and particles, such as neutrons, can fuse together to create new elements. Very high temperatures are needed for nuclear reactions to take place. Temperatures great enough existed in the very early universe and are found today inside stars, in supernova explosions, and when material falls onto a very dense star, such as a white dwarf.

All the hydrogen in the universe, and much of the helium, came into being in the first few minutes after the origin of the universe. The first stars that ever formed were almost pure hydrogen and helium. But we have seen how stars generate their energy by fusing hydrogen into helium and then later on, helium and heavier elements into a whole range, including carbon, oxygen, silicon, and iron. When a star blows up as a supernova, most of this material is blasted out into space. Even more elements are created in the heat of the explosion.

After a large number of supernovae have gone off, the material between the stars contains substantial amounts of the stuff manufactured in stars, as well as the hydrogen and helium that has been there from the start. Even the stars that do not explode contribute when they gently blow off their outer layers as stellar "winds" or as planetary nebulae.

Now we have to remember how stars form, from clouds of interstellar material. Stars being born in our Galaxy today are forming from a much richer mixture of chemical elements than the very first stars were. Not even our Sun was in the first generation of stars. It formed from a cloud with enough carbon, oxygen, silicon, iron, and so on for these elements to collect together in the swirling nebula that became the solar system and to make our planet. It is a strange thought, but most of the atoms in your own body were made in long-dead stars.

GALAXIES AND THE UNIVERSE

What is the Milky Way and when can we see it? What is a quasar? What happened in the Big Bang, and what will the universe be like in the future? Everywhere we look in the universe there are galaxies rushing away from each other. By studying them, astronomers have traced the history of the universe.

THE MILKY WAY GALAXY

The Milky Way is our home galaxy, a family of 100 billion stars. Their starlight forms a faint path of light in the night sky, parts of which are seen from everywhere on Earth. Our Galaxy has spiral arms, stars, gas, and dust. At the center there may be a gigantic black hole. A vast halo of invisible material surrounds the galactic disk.

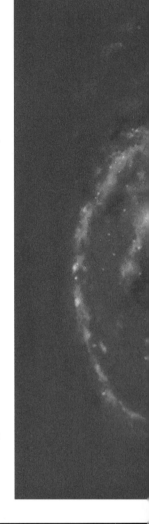

▷ The shape of the Milky Way Galaxy seen from above (left) and from the side (right). Several spiral arms wind out from the central bulge. The arms are in a thin disk containing a layer of dust. The whole Galaxy is surrounded by a spherical halo, containing globular star clusters as well as some individual stars.

To see the Milky Way, use the star charts (pages 150–153) to locate one of the prominent constellations it passes through. These constellations are Cassiopeia, Perseus, Auriga, Monoceros, Vela, Crux, Scorpius, Sagittarius, and Cygnus. Your eyes need to adjust to the darkness for at least 15 minutes.

With binoculars, fantastic clouds of stars swim into view. Now you are seeing our Galaxy just as Galileo glimpsed it four centuries ago. He was amazed to see that it consisted of crowds of faint stars, and realized this meant the heavens were far larger than anyone had ever imagined. The richest star fields lie in the southern Milky Way, and they are a splendid sight in South America, southern Africa, Australia, and New Zealand. For northern observers, the Milky Way is at its best in summer and autumn, when the constellation Cygnus is overhead in the evening.

An insider's view of our Galaxy

What exactly is the Milky Way? It is the view, from the inside, of the great starry Galaxy to which the sun belongs. There are 100 billion stars altogether, arranged in a thin disk with spiral arms. Because we live inside the Galaxy, it is hard to imagine the shape directly. When we see the Milky Way in the sky, we are looking through the plane of the disk.

Our view of the Milky Way is cut short by the intervening gas and dust clouds. Radio waves can penetrate these, and radio astronomers have shown that the galaxy is a large spiral, with the Sun located about 25,000 light-years from the center. The diameter of the main part of the disk of stars is about 100,000 light-years, but the thickness is far less. In the part near the Sun, it is just a few hundred light-years thick.

▽ This picture shows how the Milky Way Galaxy might look from an imaginary planet in orbit around a star in the halo of the Galaxy, way above the main disk with its spiral arms.

▷ This infrared picture of the Milky Way made by instruments on the COBE satellite shows up the central bulge of our Galaxy and its thin disk shape.

MILKY WAY FACT FILE
Size: 100,000 light-years across
Thickness near the Sun: 500 light-years thick in the disk
Central bulge: 20,000 light-years across, 3,000 light-years thick
Contains 100 billion stars

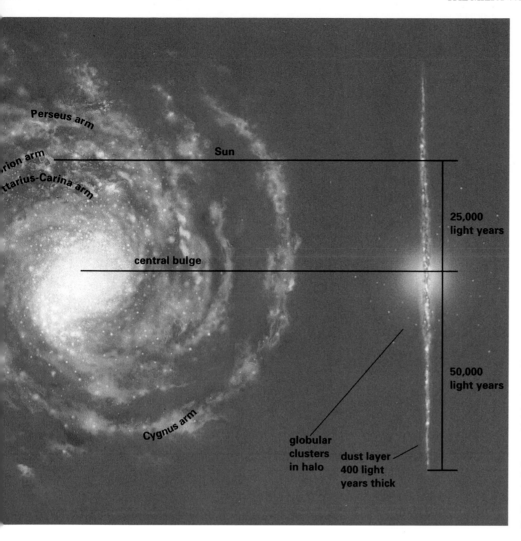

Perseus arm

Orion arm

Sagittarius-Carina arm

Sun

central bulge

25,000 light years

50,000 light years

Cygnus arm

globular clusters in halo

dust layer 400 light years thick

bright emission nebulae in the spiral arms. The open clusters, families of the youngest stars, are in the arms.

Runaway stars

Most stars in the Sun's vicinity move around the Galaxy at speeds of 20 to 30 miles per second. However, some stars travel at speeds more than twice as high as that. These speeding stars have orbits that cross right through the disk of the Galaxy. Stars out in the galactic halo have very high speeds.

The invisible Galaxy

Astronomers measure the amount of matter the Galaxy contains from the orbital speeds of stars and gas. The faster a star is orbiting at a given radius, the more massive its galaxy must be. Exactly the same method is used to find the mass of the Sun, where the orbital speed of a planet is linked to its distance from the Sun and the Sun's mass.

The Sun's speed and distance from the galactic center imply that the mass of the Galaxy within the Sun's orbit is about 100 billion solar masses. This is roughly the same as the amount of mass we can see as stars and gas.

However, stars farther out than the Sun tell a very different story. Instead of traveling more slowly at greater distances from the center (as happens with the planets in the solar system), their speeds remain more or less constant. This can happen only if the stars are pulled by a much greater gravitational force from a huge amount of invisible matter. Clusters in the galactic halo move as if pulled by about 10 times more matter than we can see.

The Milky Way has two satellite galaxies, the Large and Small Magellanic Clouds. The orbit of one of these tells us that the mass in the halo is 5 to 10 times higher than the mass we see in the disk.

Dark matter in the halo

Most of the matter in the galactic halo is invisible, so it cannot be made up of ordinary stars. Nor is it gas, because this would have been detected by radio telescopes or ultraviolet telescopes. The light

The inner part of the disk has a central bulge, a sphere of stars about 3,000 light-years thick. In this region the stars are packed much closer together than in the disk. The spiral disk with its bulge are inside a vast halo of matter that extends out to 150,000 light-years from the center.

In the disk

The disk of the Galaxy resembles a thin pancake. There are four spiral arms, and in these we find gas, dust, and young stars. Our Sun is in the Orion Arm, a branch that includes the Orion Nebula and the North American Nebula. Between the Sun and the central bulge is the Sagittarius–Carina Arm, which is about 75,000 light-years long.

The Galaxy rotates. The inner parts make an orbit much faster than the outer regions. The same is true in the solar system, where Mercury orbits in 88 days, but Pluto takes 243 years. Our Sun's

galactic circuit takes about 200 million years. The Sun is about 25 galactic years old because it has made 25 orbits around the Galaxy.

Because the regions closer to the galactic center orbit much faster, we need to understand why the spiral arms have not wrapped themselves around and around the Galaxy hundreds of times, in a cosmic whirlpool. The answer is that the spiral arms are a "density wave": they are a traffic jam on a cosmic highway in which the blockage is always there, even though individual "vehicles" (stars in the Milky Way) are passing through.

As the stars and gas orbiting inside the Galaxy approach a spiral arm, they crash into the slower-moving material in the arm. But inside the collision zone, star birth takes place. As the gas and dust piles up in the congestion, compressed clouds collapse under the force of gravity and make new stars. When we look at other spiral galaxies, we see young stars and

from distant galaxies shines right through the halo, so the extra mass is not dust. The dark matter could be atomic or nuclear particles of a kind not yet detected on Earth. On the other hand, enormous numbers of cold "planets" or black holes could account for the missing mass. Right now nine-tenths of the Milky Way Galaxy is invisible. We shall see that this missing mass problem extends to other galaxies also, as well as to the entire universe.

The center

The center of the Milky Way Galaxy is in the direction of Sagittarius. The center cannot be seen through optical telescopes because vast amounts of dust cut out the light. However, radio waves and infrared radiation can get through, and they tell us about the galactic center.

Within 1,000 light-years of the center, the stars are very crowded together. If you were on a planet in this congested region, you would see a million very bright stars in the night sky, and it would never get dark. The nearest stars would be only a few light-days away.

At the heart of the Milky Way a great deal is going on. The central region is an intense source of radio waves, infrared radiation, and X-rays. The powerful infrared emission comes from a region only 20 light-years across. Radio maps of the region show clouds of gas streaming into the center. A clumpy ring of gas twirls around the center, with hot gas falling from its inner edge into the center.

The central monster

At the heart of the Milky Way there is a mysterious source of immense energy. Shining like a hundred million suns, it is small enough to be contained entirely inside the orbit of Jupiter. The total mass is about a million times greater than the Sun's mass. This matter is almost certainly a black hole, which is greedily eating the interstellar gas and dust, pulling fresh supplies in from the clumpy ring. As this gas falls toward the black hole, it heats up and sends out the energy we now see.

Not all astronomers agree that the energy has to be made by a central black hole. They say an intense burst of star birth could have released the energy.

Our neighbors, the Magellanic Clouds

The Milky Way's two satellite galaxies, the Large and Small Magellanic Clouds, were discovered in the 16th century by Portuguese navigators sailing to southern Africa. They were later named in honor of Ferdinand Magellan (1480–1521), who led the first fleet to sail right around the Earth (1519–22). The Magellanic Clouds are visible in the Southern Hemisphere. The Large Cloud is about 165,000 light-years away, and the Small Cloud, 200,000 light-years distant.

The Large Cloud has a central bar of stars but no spiral structure. It contains about 20 billion stars and is a medium-sized galaxy. It is 10 times closer to Earth than the nearest large galaxy. Because

individual stars can be seen in the Large Cloud, it is often observed by astronomers who want to learn about the life histories of typical stars.

There is a gigantic emission nebula, called the Tarantula Nebula, in the Large Cloud. This is a huge cloud of supergiant stars and gas. It is an important "star factory." A famous supernova explosion occurred in this vicinity in 1987 (page 122).

Galactic cannibalism

Both of the Magellanic Clouds orbit our Galaxy. Because they are a long way from us, their motion across the sky is almost imperceptible. However, astronomers succeeded in measuring it in 1993, by comparing photographs taken 17 years apart. The stars in the Large Cloud had

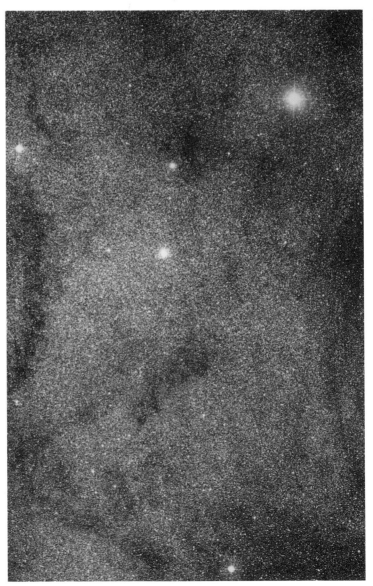

◁ Huge numbers of stars crowd this photograph of the Milky Way in the constellation Sagittarius. Here we are looking in the direction of the center of the Galaxy where stars are most densely concentrated.

moved just enough to give away its motion. From the speed, astronomers worked out the orbit of the Large Cloud. When they did so, there were two big surprises.

First of all, the speed was greater than expected. This could only be explained by assuming the Milky Way is even bigger than astronomers had previously thought. The invisible massive halo seems to extend about 10 times as far as the spiral disk of the Galaxy. The Large Cloud takes about 2.5 billion years to orbit the Milky Way.

Second, the orbit comes quite close to the massive halo. As a result, the Large Cloud is being shredded apart by gravitational forces each time it comes in close. A huge trail of debris, consisting of star clusters and hydrogen gas, is sucked away. This has created a great wispy arc of matter detached from the Large Cloud, which is now falling toward the Milky Way. The Small Cloud is suffering a similar fate. The satellite galaxies are like giant comets, on a galactic scale, leaving tails of debris behind. Astronomers calculate that within another 10 billion years the Milky Way will have grabbed all the matter in the Magellanic Clouds in an act of galactic cannibalism.

Pathway to the universe

All the stars in the Large Magellanic Cloud are at the same large distance (more or less) from us. This is a bit like saying that all the people in New York are the same distance from Los Angeles. It means that differences in the magnitudes of the Cloud stars are entirely due to differences in their ages and chemical makeup. When we look at stars in our own Galaxy, we have to correct for wide variations in distance from us, and it is difficult to measure distances accurately. For the stars in the Magellanic Clouds, it is acceptable to say there are hardly any distance effects when comparing one star with another.

In 1912 Henrietta Leavitt (1868–1921) of the Harvard College Observatory discovered more than 20 Cepheid variable stars in the Large Cloud. Four years later, she showed that the period of variation of a Cepheid depended on its luminosity. This incredibly important discovery allowed astronomers to figure out how much light the star was sending out. By relating that quantity to the apparent brightness, they could find the star's distance from our Galaxy.

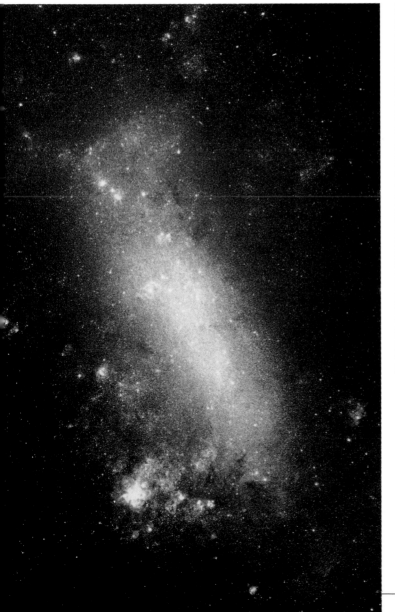

◁ △ The Large Magellanic Cloud (left) and the Small Magellanic Cloud (above) are two small, irregular galaxies that are close companions of the Milky Way. They are visible from places in the Earth's Southern Hemisphere as hazy patches in the night sky.

GALAXIES IN THE UNIVERSE

The Milky Way Galaxy has a family of neighboring galaxies, known as the Local Group, which together make a cluster of galaxies. The nearby galaxies include beautiful spirals. One of these, the Andromeda Galaxy, is the most distant object that can be seen with the naked eye. Most galaxies in the universe are either spiral or elliptical in shape, and many are members of clusters of galaxies.

During the 19th century and the early part of the 20th century, astronomers were mystified by the misty patches of light that they observed through their telescopes. Stars were obviously part of the Milky Way. So were the bright gas clouds, such as the Orion Nebula. But in their searches for comets and planets, astronomers such as Charles Messier and William Herschel had recorded seeing thousands of fainter nebulae, many of which were spirals.

Astronomers wondered whether these were galaxies far beyond the Milky Way, or clouds of gas within the Milky Way. Only by finding a way of measuring distances to the fainter nebulae could these questions be answered.

In 1924 the American astronomer Edwin Hubble showed beyond a doubt that the spiral nebulae are huge galaxies similar to the Milky Way, but at vast distances. In one stroke, he discovered the stupendous extent of the universe.

Hubble first found Cepheid variable stars in the Andromeda Galaxy. These were much fainter than the Cepheids in the Magellanic Clouds. The difference in brightness meant that the Andromeda Galaxy had to be more than 10 times farther away than the Magellanic Clouds.

The Andromeda Galaxy can be seen with the naked eye alone, and is the most distant object visible without using binoculars or a telescope. Countless galaxies are far fainter than this, and therefore are at greater distances. Edwin Hubble had found the realm of the galaxies. Within the next few years he measured the distances to many other spirals and was able to show that even the nearby galaxies are many millions of light-years away. The dimensions of the known universe far exceeded previous guesses.

Types of galaxies

There are many different kinds of galaxies. They vary in shape, size, mass, and energy output. First, we will look at normal galaxies, which derive their energy, or light, from nuclear reactions in stars. In the next chapter, we will consider active galaxies, where the energy is generated more exotically, by black holes gobbling up stars, for example.

Hubble did much pioneering work on the shapes of normal galaxies, which he divided into three shape families: ellipticals, spirals, and irregulars.

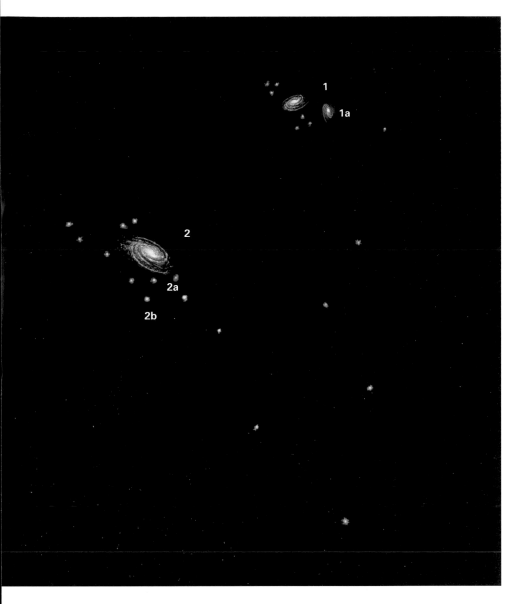

Elliptical galaxies are shaped more or less like lemons or footballs. They have no spiral arms, but they do range from almost spherical (known as E0) to a short, stubby cigar shape (known as E7). About three-quarters of all the galaxies in the universe are ellipticals. They come in a huge range of sizes, from dwarfs to supergiants. The very largest ellipticals are millions of light-years in diameter, and are the most massive galaxies known.

For spiral galaxies, astronomers make a distinction between barred spirals, which have a central bridge of stars linking the inner ends of the two spiral arms, and ordinary spirals where the arms emerge straight from the nucleus. Spirals range from about 20,000 light-years across to 100,000 light-years. The Milky Way is right at the top end of the size range.

A small percentage of galaxies do not fit into a neat shape category. These are the irregular galaxies, many of which are satellites of larger galaxies, as is the case with the two Magellanic Clouds.

No matter what the shape of a galaxy, all the stars, gas clouds, and dust within it form one family, held together by gravity. The stars within a galaxy move on orbits around and through the galaxy, which take millions of years. Everything in a galaxy is in motion, but immense periods of time would be needed to detect a significant change in their appearance.

The Local Group

As we peer into deep space, we find that galaxies are not spread uniformly all across the universe. Galaxies clump together in clusters, or families. Our own family is the Local Group. As galaxy clusters go, it is rather sparse, with about 25 members scattered across 3 million light-years. The biggest members are the Milky Way and the spirals M31 in Andromeda and M33 in Triangulum. The Milky Way has about nine dwarf galaxies moving nearby, and Andromeda another eight. Astronomers keep finding new faint galaxies in our Local Group.

Each member of the Local Group is moving within the gravity pull of all the other members. All clusters of galaxies are held together by the grasp of gravity, which is the most important long-range physical force in the universe. By

△ Our Local Group of galaxies is dominated by the Andromeda Galaxy (**1**) and our own Milky Way Galaxy (**2**), both large spirals. Both have clusters of dwarf galaxies around them (**1a**, **2a**, **2b**).

◁ No two galaxies are quite alike. These three are all spirals. The one in the center is viewed nearly edge-on and the one on the right is a barred spiral.

measuring the speeds of galaxies within the Local Group, astronomers can calculate its total mass. This is about 10 times greater than the mass of visible stars, which suggests that extensive amounts of dark matter must be present in the Local Group.

The Virgo Cluster

As we journey beyond the Local Group, we can find other small groups of galaxies, such as Stephan's Quintet, in which two spiral galaxies are meshed together. But farther out, much bigger clusters beckon. The mighty Virgo Cluster, which is about 50 million light-years away, is the nearest large cluster. It is too far away for us to work out the distance from variable stars. Instead, the magnitudes of the brightest stars and the largest star clusters are used. These brightnesses are compared to the brightnesses of similar objects whose distances are already known.

The Virgo Cluster is huge, sprawling across a patch of sky nearly 200 times the sky area of the Full Moon. This gigantic cluster has a few thousand members. Center stage are three elliptical galaxies, first listed by Charles Messier: M84, M86, and M87. They are really enormous galaxies. The biggest, M87, is similar in size to the whole of our Local Group. Virgo is so massive that its gravity not only holds together its huge membership but also extends as far as the Local Group. Our Galaxy and its companions are being dragged in the direction of the Virgo Cluster.

The Coma Cluster

Moving farther out, at a distance of about 350 million light-years, we come to a huge galactic city in the constellation Coma Berenices. This is the Coma Cluster, which has over 1,000 bright elliptical galaxies and perhaps many thousands of fainter members beyond the limits of our present vision. The cluster is 10 million light-years across, and is dominated by two supergiant elliptical galaxies at its heart. Astronomers speculate that this cluster has tens of thousands of members.

All the galaxies are locked into the cluster by gravity. In this case, the speeds of the galaxies within the cluster indicate

▷ Galaxies are concentrated toward the center in the Coma Cluster. There are hardly any spirals, perhaps because any spirals have merged to make elliptical galaxies.

▽ The spiral galaxy M33 is in the constellation Triangulum. It is one of the neighbors of the Milky Way and belongs to the Local Group of galaxies.

△ The Virgo Cluster of galaxies is the nearest really large galaxy cluster, and is at the center of a huge supercluster of galaxies. The Virgo Cluster has thousands of member galaxies.

▷ A group of five galaxies known as Stephan's Quintet, after the astronomer who first noticed them in 1877. One of the galaxies is much closer than the other four and just happens to be in the line of sight.

that only a small percentage of the total mass is in stars we can actually see. The Coma Cluster, and other rich clusters like it, is mainly composed of dark matter.

Rich clusters like Coma have hardly any spiral galaxies in their central region. This may be because the spiral galaxies that once existed have merged to form elliptical galaxies. Coma is a strong source of X-rays emitted by very hot gas, at a temperature of between 20 to 200 million degrees Fahrenheit. This gas is found in the central region of the cluster and its chemical makeup is like the matter in stars.

What may have happened is that galaxies in the central part of the cluster crashed together and were stripped of their gas clouds as they rushed about inside the cluster. The gas is heated by friction as the galaxies race through it at speeds of up to thousands of miles per second. Because the galaxies have lost their gas, the spiral arms of the galaxies have gradually disappeared.

Superclusters and voids

Deep-space photography shows the galaxies going on and on as we probe deeper into the universe. Almost wherever we look, there are tiny galaxies scattered like dust. There are objects as far as 10 billion light-years away. These countless galaxies each contain thousands of millions of stars. Numbers like these are truly mind-boggling, even to professional astronomers. The extragalactic universe is vaster than anything we can easily comprehend.

Almost all galaxies are found in clusters, which contain anything from a handful of galaxies to many thousands. But what about these clusters: are they grouped into families too? Yes they are!

The local cluster of clusters, known as the Local Supercluster, is a flattened family that contains the Local Group and the Virgo Cluster. The center of mass is in the Virgo Cluster, and we are on the outskirts. Astronomers are putting a lot of effort into mapping the Local Supercluster to find its structure in three dimensions. It includes nearly 400 separate galaxy clusters, assembled into sheets and ribbons of matter with gaps in between.

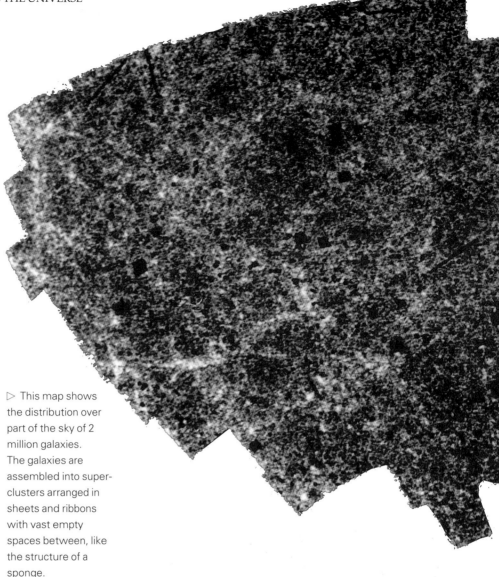

▷ This map shows the distribution over part of the sky of 2 million galaxies. The galaxies are assembled into super-clusters arranged in sheets and ribbons with vast empty spaces between, like the structure of a sponge.

There is another supercluster in the constellation Hercules. This is about 700 million light-years away, beyond a gap 300 million light-years across in which there are hardly any galaxies at all.

What astronomers are finding is that the superclusters are separated by huge volumes of space that are empty. Within the supercluster there are spherical bubbles, millions of light-years across, with no galaxies. The clusters are arranged in filaments and ribbons, giving the universe on the grandest scale a spongy structure.

Hubble's Law and redshifts

We now know that our universe is expanding, getting bigger and bigger all the time. Hubble played the major role in this discovery. He had measured the distances to nearby galaxies from the Cepheid stars, and the speeds from the redshifts. His next finding came by making a graph of the speeds of galaxies plotted against their distances. This showed that the distance and the speed of a galaxy are tied together on a straight-line graph: the farther a galaxy is from us, the greater its speed (see page 47).

The Hubble Law says that the faster a galaxy is speeding, the farther away it is. Hubble found a common link between two quantities that can be measured for nearby galaxies: the distance and the redshift (which gives the speed). Once this link had been found, the Hubble Law could be turned around. By finding the redshift for a more distant galaxy, astronomers work out its distance using

the Hubble Law. This is how astronomers find the distances to galaxies that are far out in the universe.

Of course there are uncertainties in using the Hubble Law. For example, if we make mistakes in finding the distances to nearby galaxies, the graph will not be quite right: any errors will extend out to deep space when we try to deduce the locations of the farthest galaxies.

Nevertheless, the Hubble Law is a very important method of finding out about the large-scale structure of the universe.

The expanding universe

Why does Hubble's Law show that the universe is expanding?

All the galaxies are rushing away from us. Surely this shows that the Milky Way is at the center of the universe? When we see an explosion, such as a firework exploding in the sky, we see everything fly away from the explosion itself. So, if everything is racing away from us, must we be at the center of expansion? No, we are not at the center.

As the fragments of an explosion fly apart, the distances between all the bits of debris increase. This means that any piece of the debris will "see" all the other fragments flying away.

To see how this works, get a balloon and mark some galaxies on it, using spiral and elliptical patterns. Now blow up the balloon slowly. As the balloon expands, so the galaxies get farther apart. No matter which galaxy on the

balloon you take for your main reference, all other galaxies are creeping farther away as you inflate the balloon.

We can think about this mathematically. The skin of the balloon is a curved surface: it has almost no thickness. As you blow up the balloon, this spherical surface is expanding into space, so the curved surface of the balloon grows from two-dimensional space into three-dimensional space—and while that happens the galaxies marked on the surface get farther apart.

In the case of the universe, the three dimensions of ordinary space are expanding in a special four-dimensional space called space-time. The extra dimension is time. As time passes, the three dimensions of space have a larger and larger extent. The clusters of galaxies are frozen onto the expanding framework of space, getting farther apart all the time.

The age of the universe

How can astronomers measure the age of the universe? We can find out how old a tree is by counting the number of rings in the wood, because one ring is added every year. Geologists can estimate the ages of rocks that are laid down in sediments from the fossils they contain. The age of the Moon was found by seeing how much radioactivity remained in the rocks containing radioactive elements. All these methods involve getting data, such as the number of rings or the species of fossils or the amount of radiation still emitted, and calculating an age from that observation.

To find the age of the expanding universe, we look at the distances and speeds of a large sample of galaxies. This tells us that for every million light-years we journey into the universe, the speed of the galaxies increases by about 12 miles per second (astronomers are uncertain about the exact value to within 1 to 2 miles per second). With this relative change of speed with distance, we can work out that 17 billion years ago everything was in the same location. This is one way of estimating the age of the universe, because it is the time that has passed since the Big Bang when expansion first started.

ACTIVE GALAXIES

Violent explosions occur in the centers of some galaxies. Jets of matter shoot out of their central regions, and they emit far more energy than the stars in the galaxy can make by normal means. Active galaxies may be powered by black holes in their central regions.

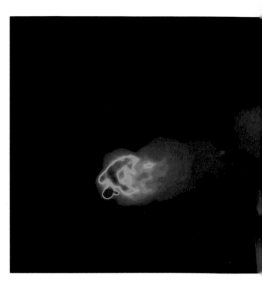

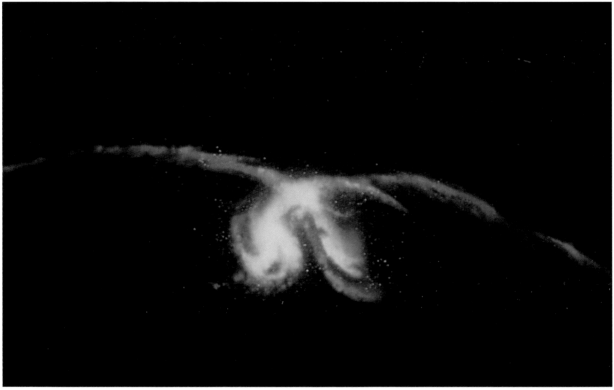

◁ A close encounter between two galaxies, as imagined by an artist. In such an encounter, gas and dust are transferred from one galaxy to the other. Some of the material may fall into a central black hole, producing huge amounts of energy.

Astronomers have found that some galaxies, just a small percentage of all the known galaxies, are extremely powerful. Normal galaxies, the vast majority of those in the universe, emit energy that comes from their stars: the light of a normal galaxy is, basically, the starlight from the billions of stars within it. For active galaxies, this is not the case. Much of an active galaxy's energy is not starlight, but something else.

How can we say that a galaxy has a source of energy in addition to that of its normal stars? Normal stars emit light because they are hot, and they have a "thermal" spectrum. The spectrum of an active galaxy is not like the spectrum of a star. Active galaxies have a lot of radiation coming from something other than hot stars. An active galaxy typically has much more infrared, radio, ultraviolet, and X-ray emission than a normal galaxy. Of course, normal galaxies may also have small amounts of these radiations. But the point is that in an active galaxy, the radio waves, or the ultraviolet emission, or the X-rays are the main form of the energy. Furthermore, the amount of energy we detect can vary a great deal over a period of just a few days.

The energetic activity in most active galaxies comes from the center, or nucleus, of the galaxy, which can be a billion times brighter than the Sun.

Radio galaxies

There are many types of active galaxies, and the first examples were found by radio astronomers in the 1950s. Radio astronomy was then a new science, having started in about 1946. At first radio astronomers believed that their crude telescopes were picking up radio waves from stars. But in 1951 at Cambridge University

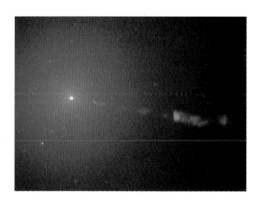

△ Cygnus A is one of the strongest radio sources in the sky. In the radio map at left, the galaxy itself would be a small dot midway between the huge radio clouds that it has blasted out into space. The photograph at right shows the central galaxy in ordinary light.

◁ △ The giant elliptical galaxy M87 (left) probably has a supermassive black hole at its center. The infrared image from the Hubble Space Telescope (above) shows the core of the galaxy as a bright point and the jet of material being hurled outward. The jet is also a strong radio source, as seen in the radio map (below left).

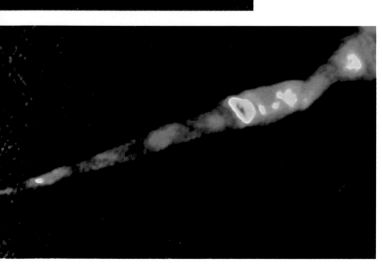

in England, Martin Ryle and Graham Smith tied down the location of one of the strongest sources of radio noise, Cygnus A. Working at Mount Palomar, California, astronomers found a fuzzy galaxy at the location of the Cygnus A radio source. They measured the redshift and used Hubble's Law to calculate a distance.

The odd-looking radio galaxy turned out to be amazingly far away. It is about 1,000 million light-years from us, 500 times farther than the Andromeda Galaxy. This fantastic distance completely astounded astronomers, because the radio signal from the galaxy was not that much different from the Sun's radio signal. Its signals are 10 million times stronger than the radio waves from Andromeda.

Think about how spread out the radio waves become as they fan out across the huge distance from Cygnus A to Earth. For Cygnus A to be such a strong radio source, located so far away, there must be a very unusual and powerful source of energy in its center. It is one of the most powerful radio galaxies in the universe.

Cygnus A has two clouds of radio emission spanning the central galaxy. In radio galaxies of this kind, the radio clouds stretch across millions of light-years.

Galaxy M87

At the center of the Virgo Cluster, at a distance of about 50 million light-years from Earth, the giant galaxy M87 beams out X-rays at a rate equivalent to a billion suns. Spurting out of the core of this supergiant galaxy is a wiggly jet of matter about 6,000 light-years long. This sends out as much light as 10 million suns and it consists of a series of blobs, each a few tens of light-years across.

Radio astronomers know M87 as Virgo A, the strongest radio source in the Virgo constellation. Radio maps show a narrow beam of energy jetting out from the center of M87. This coincides with a jet seen with optical telescopes.

The active central region of M87 is only 45 light-days across, according to very detailed maps made by radio astronomy. The orbits of stars in this region are controlled by a large concentration of mass at the center of the galaxy. It could be as much as 5 billion solar masses, all in the central part of the galaxy.

Galaxy Centaurus A

In the Southern Hemisphere there is an active galaxy just 10 million light-years away. This is Centaurus A, an elliptical galaxy crossed by a band of dust. Two large clouds of radio emission sit on either side of the visible galaxy. There is a radio jet and an X-ray jet pouring energy away from the active nucleus.

Seyfert galaxies

Nearly all the active radio galaxies are elliptical or irregular. However, some of the nearby active spiral galaxies have tiny bright spots in their centers. These are Seyfert galaxies, and about 600 are now known. They are spiral galaxies with bright and active nuclei. In their central regions are large clouds of hot gas. The heating of the gas is caused by energy streaming out of the nucleus.

A hot galaxy cluster

One of the most famous active galaxies is NGC1275, at the heart of the Perseus Cluster of galaxies, about 180 million light-years away. This galaxy cluster measures some 250,000 light-years in extent, with its members strung out across space like beads on a necklace. The radio source within NGC1275 is 1,000 times stronger than that on the Milky Way. A glow of X-rays extends throughout the cluster. These X-rays come from glowing gas that has been heated to a temperature of millions of degrees. Dense streamers of this hot gas are tumbling toward the central galaxy in the cluster.

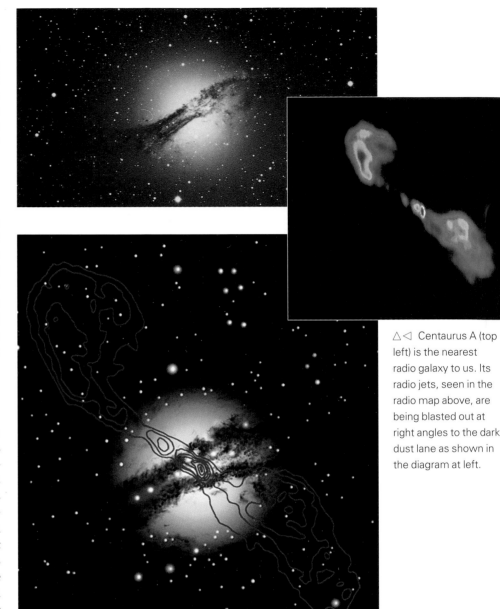

△◁ Centaurus A (top left) is the nearest radio galaxy to us. Its radio jets, seen in the radio map above, are being blasted out at right angles to the dark dust lane as shown in the diagram at left.

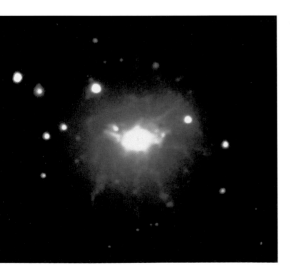

◁ ▷ The Seyfert galaxy NGC 1275 (left) is a strong source of radio waves and X-rays. Material is exploding out from it at more than 1,000 miles per second. It lies in the Perseus Cluster of galaxies (right) and is also known as Perseus A.

Quasars

Seyfert galaxies are relatively close to us, and most radio galaxies are in the middle range. Much farther away in space we find quasars, and these are extremely energetic. The discovery of quasars involved careful detective work.

The story begins back in 1960. Radio astronomers improved methods for finding out exactly where in the sky radio sources were located. Radio source 3C 48 seemed to coincide with a star unlike any other: it had bright lines in its spectrum that could not be matched with any known atoms. Then, in 1962, another star appeared to match up with a different radio source, 3C 273.

A year later, at Mount Palomar, California, the astronomer Maarten Schmidt showed that its strange spectrum could be matched to hydrogen gas if light from the starlike object was redshifted by about 16 percent. At the time this redshift was large compared to those of most galaxies. Instead of 3C 273 being an exotic star in the Milky Way, it turned out to be rushing away at 16 percent of the speed of light. Other starlike radio sources, such as 3C 48, also turned out to have high redshifts. These high-redshift compact objects, which are starlike on photographs, are quasars.

The word *quasar* was invented as a short form of *quasi-stellar* radio source. *Quasi-stellar* means "looking like a star but not really a star." Astronomers now believe that quasars are the brightest kind of active nuclei seen in galaxies. Thousands of quasars have been identified.

Although the first few were found by radio astronomers, only one-tenth of those now known emit radio waves. They look like stars when photographed (which means they are small compared to galaxies), but they have large redshifts. The highest redshifts are nearly 5. For these the wavelength of the light sent out by the quasar has been stretched nearly sixfold. This distortion is very much greater than that of most galaxies, although the largest telescopes are now finding a few extremely faint galaxies with high redshifts.

The light from distant quasars has taken billions of years to reach us, so they tell us about conditions in the universe long ago.

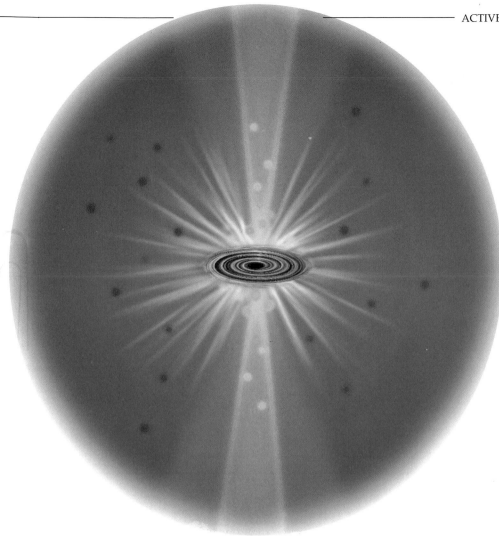

△ Quasars are embedded in galaxies. In almost all cases, however, the quasar shines so brightly that it swamps the much fainter light of its parent galaxy. Consequently, only the point of light from the active nucleus shows up on photographs. At the middle of the quasar there is an extremely powerful source of energy, almost certainly a black hole. This is surrounded by a disk of material a few light-years in diameter. Close to this disk there are rapidly moving gas clouds and farther out, at about 100 light-years, thinner, cooler clouds, where the quasar merges with the parent galaxy.

NAMING GALAXIES

Many of the brighter normal galaxies have familiar names, such as the Andromeda Galaxy and the Magellanic Clouds. The active galaxies often have names that seem like code numbers. This is because they were first detected in surveys of the sky and then published as lists of new objects.

Over 200 years ago, Charles Messier produced lists of objects that could be confused with comets when viewed through a telescope. His 87th object, now known as M87, is a powerful active galaxy at the center of the Virgo cluster of galaxies. Martin Ryle, the Cambridge radio astronomer, listed 471 radio sources in 1959. This was the third survey of radio sources made by Cambridge University, and the objects listed are known by their 3C number: Cygnus A is 3C 405. Today there are such huge numbers of objects in the surveys made by astronomers that they have resorted to using numbers related to the object's position, and these really do look like secret codes! For example, one of the most distant objects in the universe is known as 10214+4724.

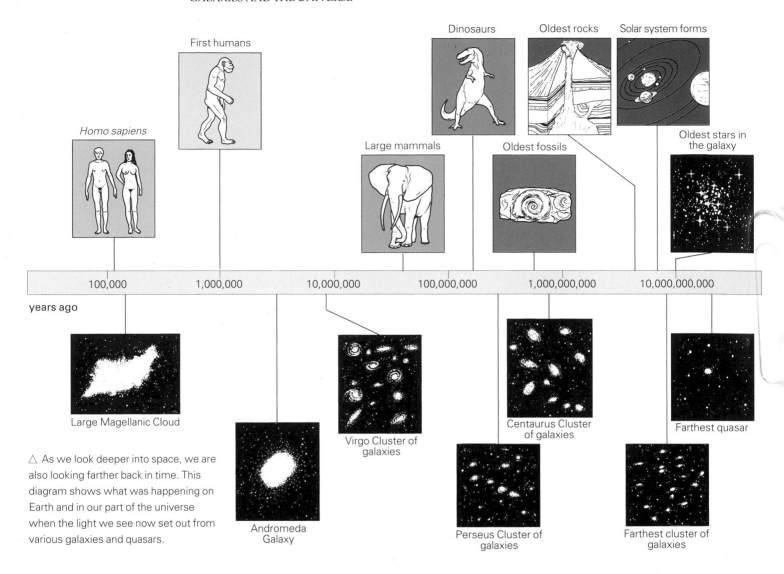

First humans

Homo sapiens

Dinosaurs

Oldest rocks

Solar system forms

Large mammals

Oldest fossils

Oldest stars in the galaxy

100,000   1,000,000   10,000,000   100,000,000   1,000,000,000   10,000,000,000

years ago

Large Magellanic Cloud

Virgo Cluster of galaxies

Centaurus Cluster of galaxies

Farthest quasar

Andromeda Galaxy

Perseus Cluster of galaxies

Farthest cluster of galaxies

△ As we look deeper into space, we are also looking farther back in time. This diagram shows what was happening on Earth and in our part of the universe when the light we see now set out from various galaxies and quasars.

Where are quasars located?

Most quasars have very high redshifts. Edwin Hubble showed how the redshifts of galaxies can be used to find their distances. Can we do the same for quasars? In other words, does the redshift of a quasar tell us its distance? Many astronomers think that quasars probably do follow Hubble's Law.

The large redshifts of quasars mean that they are very far away—billions of light-years, in fact. The most distant quasars are 10,000 million or more light-years from us. Quasars are important in astronomy for two reasons. First, they have to be incredibly energetic for our telescopes to be able to pick them up at such vast distances. Second, since their light has taken billions of years to reach us, they can tell us about conditions in the universe long ago. Astronomers want to know what makes

quasars shine so brightly, and by observing the most distant quasars they can see what the universe was like long before our Sun existed.

Viewing the active centers

Active galaxies and quasars produce much more energy than normal galaxies, which is why we can see them at such great distances. In ordinary galaxies nearly all the light comes from normal stars. In energetic galaxies the total output of energy far exceeds the amount made by stars. Very detailed maps made by radio astronomers have shown that most of the extra energy comes from the central regions of galaxies.

With the Very Long Baseline Array (VLBA), a huge American radio telescope, astronomers can view the active centers of powerful galaxies. This telescope has 10

dishes, spread across the continent and over the Pacific west to Hawaii. With a baseline of 5,000 miles, it can peer into regions just a few light-days across. The Hubble Space Telescope can see great disks of hot gas swirling around the nuclear regions of a few active galaxies. Together, these modern telescopes provide glimpses of the central engines that power up the active galaxies.

Black holes in galaxies

The cores of energetic galaxies are now widely believed to harbor gigantic black holes. These probably range in size from a few thousand to a few billion times the mass of our sun. The Hubble Space Telescope has seen whirlpools of matter rotating around the black holes. Once a black hole has formed, it grows by pulling in more matter from the surrounding

◁ △ The Hubble Space Telescope made this close-up image (above) of the core of the spiral galaxy M51 (left). The dark cross is due to the absorption of dust. The darkest bar is thought to be a dust ring 100 light-years across surrounding a black hole.

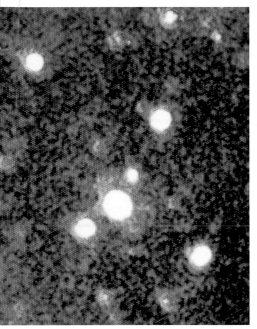

◁ The quasar 3C 275.1 is the brightest object near the center of this picture. The quasar nucleus is surrounded by a rotating elliptical gas cloud. 3C 275.1 is 7 billion light-years away. The light received from it now set out 2 billion years before the solar system formed.

△ Huge amounts of energy are generated as material falls into a black hole. The black hole beams away its energy in jets fired out along its rotation axis.

◁ One of the radio dishes of the Very Long Baseline Array telescope, in Pie Town, Texas.

regions. This increases its gravitational pull, enabling it to suck in yet more material. In giant galaxies such as M87, the central black hole may actually be gobbling up the equivalent of several stars a day.

With the VLBA, the anatomy of these central engines in active galaxies can be explored further. The aim is to see if huge black holes drive the energy flows. As matter falls toward a black hole, it is first forced into orbit around the black hole. This is because the incoming matter must first lose some of its energy of motion. The result is that the black hole is shrouded by a rotating disk of incoming matter. At the inner edge of the disk, the matter finally cascades down the throat of the black hole. Close to the hole itself, but still outside it, stupendous amounts of energy are beamed out in jets along the rotation axis of the black hole. Such jets are seen by optical and radio telescopes in objects such as M87.

The black hole and its surrounding disk are constantly refueled with more matter. The central regions of galaxies are packed with stars. Very dense star clusters may supply the fuel. This could be gas that is wafted away from the surfaces of normal stars during their evolution, or it could be the debris from a very high number of supernova explosions. As the black hole gets more massive, the growing strength of its gravitational field makes it easier for it to capture stars and shred them apart.

In normal stars, energy is released when hydrogen is changed to helium by nuclear fusion. This process converts less than 1 percent of the mass to energy. A rotating black hole is much more efficient than this. In theory, almost half the infalling matter can be converted to energy. Certainly a much higher proportion of the mass is changed to energy. For the most energetic galaxies in the universe, most of the energy is probably being released by rotating black holes rather than by nuclear burning inside normal stars.

QUASARS

Quasars are the most distant objects seen through telescopes. Some quasars are 15 billion light-years away. Light from a very distant quasar can be bent by gravity if it passes through a galaxy cluster on its way to Earth.

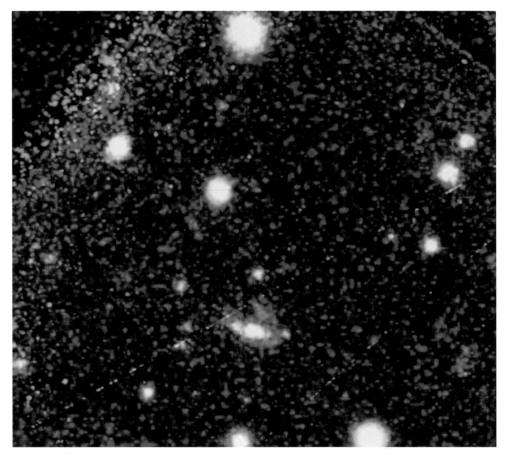

△ ◁ Some of the most remote known objects in the universe. Above top: a cluster of galaxies at least 7 billion light-years away found by the Hubble Space Telescope. Above bottom: the most distant known quasar, PC1247+34. Its redshift is 4.9. Left: 4C41.17, the most distant known galaxy in the universe. Its redshift is 3.8, and it is estimated to be 12 billion light-years away.

Thousands and thousands of quasars are now known, almost all of them several billion light-years from our Galaxy. The most distant quasars are traveling away from us at more than nine-tenths of the speed of light. To find very remote objects, astronomers make surveys of large numbers of faint objects. A large optical telescope can now obtain the spectrum of many hundreds of objects a night, thus speeding up the search for high-redshift quasars.

Very distant objects allow astronomers to travel in time. When we see a star or galaxy that is 10 billion light-years away, we are looking at something that is 10 billion years younger than our Galaxy at the time when we observe it. This is because the light has taken 10 billion years to reach us. No doubt the distant galaxies have changed greatly over billions of years.

By looking at remote galaxies, astronomers can do something that is impossible for historians: astronomers can actually look back to the early universe and see directly what conditions were like in the past, whereas historians have to use incomplete evidence that has survived from the past.

Part of the reason why larger and more efficient telescopes are needed is so that we can learn more about what the universe was like in the past by observing its most distant members. We see these objects at a time shortly after galaxies first started to form.

Gravity makes a lens

Einstein's theory of gravity says that light is bent if it goes through a strong gravitational field. A famous test of the theory was carried out at a solar eclipse in 1919. The positions of stars seen close to the Sun were altered slightly because rays of light were deflected as they passed very close to the edge of the sun.

Quasars can show this effect as well, but much more dramatically. Quasars are rarely found very near each other in the sky. But in 1979 astronomers found a pair of identical quasars very close together. These are actually a pair of images of a single object whose light has been distorted by a gravitational lens. A small but very dense mass lies somewhere along the line of sight to this quasar. Its gravity splits the light into a double image.

Many more gravitational lenses are now known. Some of them produce multiple images of distant quasars. In other cases, the far-off quasar is blurred out into a beautiful arc of light. The way in which the illusion is created is that the light from the distant quasars passes through a galaxy cluster on its way to Earth. If the galaxy cluster has a dense concentration of mass, perhaps a giant black hole or a huge elliptical galaxy, then a distorted image results.

In one case, the alignment of a quasar with an intervening mass was so good that the radio image of the quasar was a near-perfect ring.

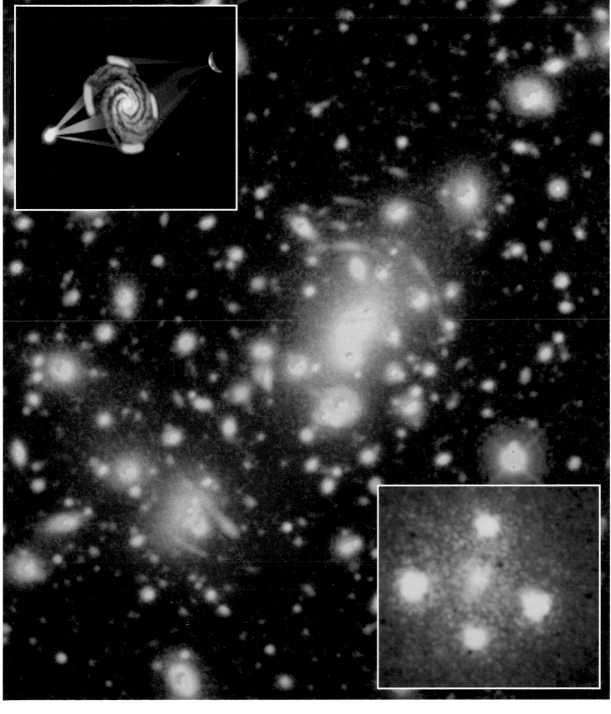

◁ The main photograph shows a giant cluster of galaxies (colored yellow) about 2 billion light-years away, and more than 30 more distant galaxies (mainly blue) distorted into arc shapes by the strong gravity of the giant galaxy cluster. The inset at top left shows how a galaxy acting as a gravitational lens distorts the image of a more distant object. At the lower right is the multiple image of a quasar, known as the Einstein Cross.

THE UNIVERSE FROM BEGINNING TO END

Our universe is extremely large, and the most distant galaxies seen within the universe are about 15 billion light-years away. The remote objects help us to discover the history of the universe, which began in a stupendous explosion. The future of the universe is uncertain: it could expand without limit or eventually collapse under its own gravity.

To understand the universe as a whole, astronomers make certain assumptions. We cannot carry out any experiments on the universe as such. All we can do is observe its contents and then use the tools of science to see what the universe is prepared to tell us about its origin and its future. To use science as a tool for exploring the universe as a whole, we take three things for granted.

First, we assume that the laws of science are the same everywhere in the universe. This means that the law of gravity we have tested here on Earth applies to the entire universe. It means that the properties of atoms and molecules are the same wherever you look in the cosmos. Without this assumption that physics and chemistry and mathematics "work" in the same way throughout space, we could not make heads or tails of the universe.

Second, we assume that the universe is spread out more or less uniformly throughout space. True, there are lumps and clumps and voids, which we see as galaxy clusters. But on the grandest scale, the assumption is that matter and radiation are smoothly distributed.

Finally, we make an assumption about the geometry of space itself. Space is assumed to have the same properties in all directions. There is no special direction in space and furthermore the universe has no center and no edge.

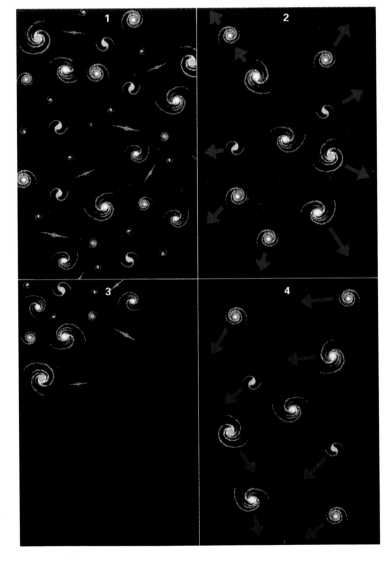

◁ When trying to understand the universe, astronomers assume that it is homogeneous (**1**)—meaning that matter and radiation are spread out more or less uniformly; and isotropic (**2**)—meaning that space is the same in whatever direction we look. In an inhomogeneous (non-homogeneous) universe (**3**), matter and radiation would not be uniform. In an anisotropic (non-isotropic) universe (**4**), the expansion of the universe would be different in different places.

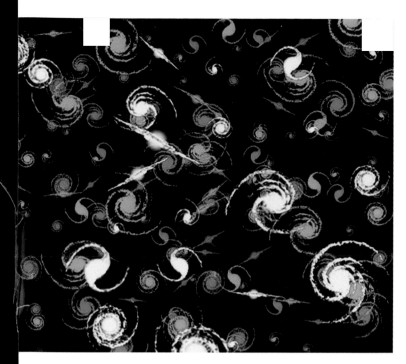

◁ Because the galaxies are all rushing apart, their light is weakened by the time it reaches Earth. The most distant ones (looking smallest) are receding the fastest and so are dimmed the most. This is why the night sky does not glow as bright as day.

Cosmology

There are many theories about the nature of the universe. The Ancient Egyptians, the Greeks, and the Romans all believed that gods of various kinds controlled the cosmos. Scholars in the Middle Ages believed that the universe was quite small, with Earth at the center. Surrounding everything were the heavens, the realm of God and the angels. From the time of Newton onward, however, scientists have attempted to make models of the universe in which mathematics and physics, rather than religious belief or myth or tradition, are used to explain how the physical universe works. The branch of science in which the universe as a whole is studied and modeled is called cosmology.

Cosmologists hope to describe what the universe is like on the biggest scale, what conditions were like in the past, and how the universe changes with time.

The curved universe

If you could shine a torchlight far into the universe, the beam of light would be curved. This is because matter in the universe causes itself to be curved. The situation in the universe is similar to that on the surface of the earth. Locally, the Earth seems flat, but on the large scale it is curved. You can travel around the Earth as much as you like and go in any direction, but you cannot find an edge to the surface. In theory, you might be able to make a journey right around the universe and loop back on yourself, using the curvature of space.

Scientists are not sure what the curvature of space is really like. If it is like the spherical curvature of the Earth, then the universe is "closed." This means that the universe is limited in extent, although it still has no boundary. On the other hand, maybe the surfaces curve outward, like a horse's saddle. This is trickier to understand, but a universe like that is "open" because the curved surfaces go on extending without ever wrapping back on themselves. A universe with this sort of curvature will last forever and extends without limit.

Observations of very distant galaxies and quasars should be able to help us decide what the curvature of space is like. This is one of the reasons why astronomers are anxious to build large telescopes that should answer this major question.

The dark night sky

One very simple observation gives us a clue about the nature of the universe. The sky is dark at night!

When the Sun has set, the stars give only a feeble glimmer of light, not enough even to cast a shadow. True, the stars are far away, so each star is quite dim. In the 18th century some astronomers realized that the darkness of the night sky is an important fact that needs explaining. The German Heinrich Olbers (1758–1840) wrote a scientific report on this problem in 1823.

Olbers and others thought about what the night sky would be like if the universe had always existed, extended infinitely in every direction, and did not change with time. In that case, anywhere you looked in the sky you would be seeing a star somewhere, even if it was extremely far away. Although many of the stars would be very far away, that does not alter the fact that every line of sight should end on a star's surface. We cannot see distant stars individually. However, we can see the Milky Way even though most of its stars are too faint to be seen separately. So, if every line of sight ends on a star somewhere, the sky should always be a blaze of light. So why is the sky dark at night?

The explanation is that the assumptions made about the universe were wrong. The universe and the objects in it do change and evolve with time, and the universe as a whole is expanding. This means that most of our theoretical sight lines do not land on a star. But the important point about the darkness of the night sky is that it does tell us something about the history of the universe.

The expansion of the universe

One of the most important observations in cosmology is that the clusters of galaxies in the universe are traveling away from us and from each other. The distances between the great galaxy clusters are increasing steadily with time and the universe is growing ever larger. This discovery, made by Hubble about 70 years ago, is one of the great findings of modern astronomy.

What we see today is that the clusters of galaxies are being separated by greater and greater distances as time passes. This is the expansion of the universe. As time goes by, the universe gets ever larger. Furthermore, it was denser in the past than it is now.

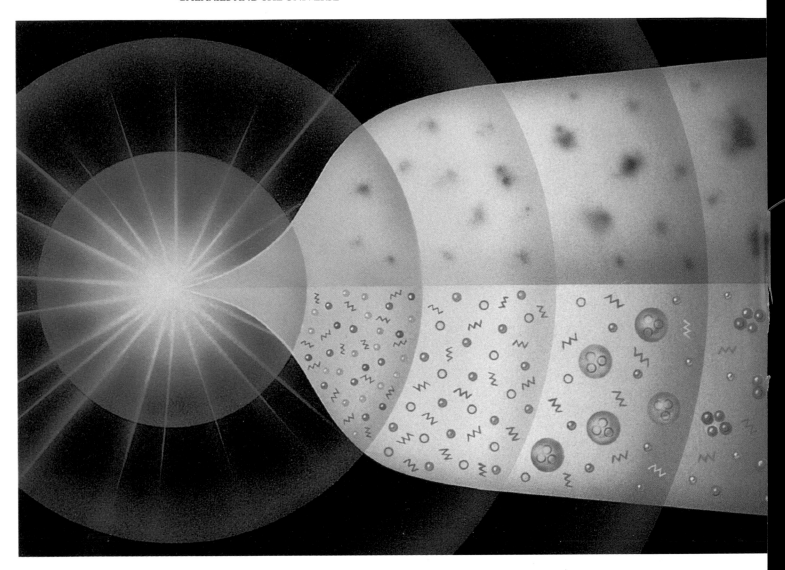

Suppose we could make a video of the history of the universe. It would show the galaxies rushing ever farther outward. But if we rewound the video, we would see the galaxies tearing toward us. Eventually, they would smash back together.

At the very beginning of the universe, all matter and all radiation were concentrated into a tiny region of space, much much smaller than the nucleus of a single atom.

The theory that the universe has expanded from next to nothing into its current immensity relies on the assumption that the redshifts of galaxies are caused by motion away from us.

The Big Bang

The universe is expanding, so it must have been denser in the past. How dense? One model of the universe, the Big Bang theory, says that in the past all matter and energy in the universe was packed into a tiny region of space. It was incredibly hot and fantastically dense. The universe we see emerged explosively from this hot Big Bang. Ever since the initial explosion, the universe has expanded and cooled.

The age of the universe cannot be calculated precisely because it depends on the amount of curvature in space, and that has not been measured accurately. From the rate of expansion, it seems that the universe is now 12 to 20 billion years old. There are great uncertainties, but the universe is definitely much older than the Sun and the Earth. The universe in the remote past, as it emerged from the Big Bang, was very different from the universe of planets, stars, and galaxies that we have today.

The first few seconds

The very early universe was a fireball of radiation. Matter was not like the stuff we see around us today. The universe consisted of a mixture of exotic particles, cooling rapidly as the tiny universe expanded. By the time the universe was about a millionth of a second old, much of the energy had been converted into protons (the nuclei of hydrogen atoms). In the next millisecond electrons formed, and these collided with protons to make neutrons. Neutrons survive only a thousand seconds as independent particles, so the next few minutes were crucial!

During the first quarter of an hour, the protons reacted with neutrons, which were fast decaying, to make the nuclei of helium atoms. In a race against time, as the universe continued to cool and expand, the universe managed to convert about a

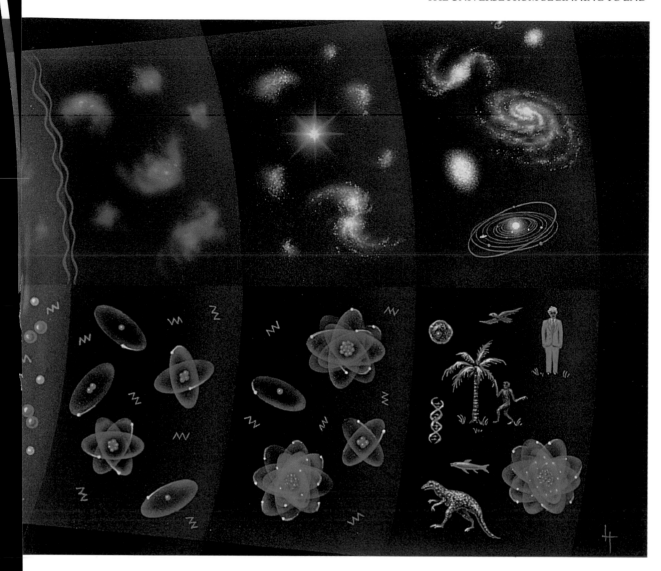

◁ Most astronomers think the universe began with an "event" about 17 billion years ago, called the Big Bang. This diagram represents, from left to right, the changes in the universe over time from the Big Bang to the present day. The upper section shows the hot, dense clumps of matter that eventually became galaxies. The lower section at first shows radiation and subatomic particles. The particles came together to make atoms and eventually, on Earth, living plants and animals.

quarter of its matter from hydrogen into helium. The remaining hydrogen was used to make the stars.

A million years go by

At the end of the first hour or so of the Big Bang, the universe consisted of photons of radiation, together with electrons, hydrogen nuclei (protons), and helium nuclei. There were no atoms because the temperature was too high for electrons to orbit the protons or helium nuclei. Any electrons that tried to do so were kicked away in collisions with energetic photons.

But time was against the radiation. Continuing expansion cooled the universe and made the photons less and less energetic as they had to fill ever-expanding space. After nearly a million years, the temperature had dropped back to 7,000°F, which is cool enough for nuclei

to hold on to any electrons that attach themselves in orbits.

It was at this stage in the life of the universe that atoms formed. It took a few thousand years for the electrons to get attached to protons or helium nuclei. Until this happened, light could not travel very far in the universe. If you had lived in the early universe, you would have been surrounded by hot bright fog, unable to see even an inch through the stuff. Once stable atoms formed, the universe became transparent and light could travel unimpeded. With our telescopes we cannot see back in time any further than this because at all earlier times the universe was opaque. Just as we cannot see the interior of the sun because it is opaque, so our telescopes cannot let us see the original fireball. We can only deduce what conditions were like by using the laws of physics and applying them to the expanding universe.

The background radiation

The heat radiation from the Big Bang is still around in the universe. In 1964 two scientists at Bell Laboratories in New Jersey were working on an annoying source of radio waves that was interfering with communication. Using a sensitive antenna, they found that the radio noise did not vary with time of day or direction in the sky. By accident, Arno Penzias and Robert Wilson had discovered radiation left over from the Big Bang.

In 1989, NASA launched the Cosmic Background Explorer, or COBE, to study the background radiation. This satellite carried liquid helium to cool its detectors down to about the same temperature as the background radiation. One of COBE's great successes was to show that the faint radiation is exactly like radiation from a warm object. This was a crucial finding because it proved that the radio waves were a form of heat radiation.

To an astonishing extent, the background radiation is the same from all parts of the sky. This means that the universe is the same in all directions, so the COBE results confirmed one of the assumptions made by cosmologists. The smooth distribution of the radiation is strong evidence that we are seeing the heat content of the universe itself, rather than contributions from large numbers of very distant objects.

The universe inflates

Different parts of the universe are very similar, to a much greater extent than astronomers would once have guessed. Opposite parts of the sky have more or less identical properties. This is a puzzle.

Although we can see two objects on opposite sides of our sky, they are in fact separated from each other by many times the distance light could have traveled since the beginning of the universe. Objects we can see cannot necessarily "see" each other. So how has the universe arranged itself to look the same all over?

Cosmologists have suggested a way of explaining how parts of the universe that are out of reach of each other today may once have been in contact. They say the universe may have expanded at a stupendous rate, far faster than the speed of light, very early on. In this theory, the whole of the universe we see today was once packed into a region of space smaller than a proton. In a single "whoosh," this ballooned up into something millions of miles across. This all happened in the first millionth of a millionth of a millionth of a millionth of a millionth of a second (a zillionth!). The inflationary universe theory explains why the universe today is so large and so uniform: it got spread out tremendously fast very early on.

If the theory of the inflationary universe is correct, the universe we see through our telescopes is only a small fraction of the total cosmos. Most of the universe is so far away that we cannot see it because the light has not had enough time to reach us.

The speeding Milky Way

There is a slight redshift in the background radiation in one area of the sky, and a slight blueshift in the opposite area. This is because the Milky Way Galaxy and other nearby galaxies are cruising through the universe. The Local Group of galaxies has a speed of about 350 miles per second relative to the distant galaxies and the large-scale universe. Perhaps a large concentration of mass, so far undiscovered, is dragging the Local Group by the force of its gravity. There is a large concentration of galaxies, known as The Great Attractor, that could be doing this pulling.

The formation of galaxies

One of the great issues in astronomy is to find out how the structure of the universe formed. The background radiation shows that the early universe was quite smooth. As the expansion raced along, how did the universe manage to form galaxies before all the matter became too spread out? Ground-based telescopes and the Hubble Space Telescope are providing the first direct clues from the time when galaxies were formed.

Galaxies like the Milky Way are probably between 10 and 12 billion years old. Since their formation, they have changed considerably. For example, billions of years of stellar evolution have changed the chemical makeup of the local galaxies. To find out what galaxies were like when they first formed, we can look out at the distant universe, where the galaxies are much younger.

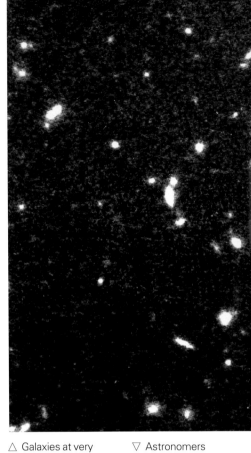

△ Galaxies at very great distances in the universe are seen as they were long ago, when the universe was younger. Many photographs of remote galaxies appear to show galaxies merging into each other. Perhaps galaxies were smaller when they first formed, but many have grown much larger as a result of capturing and merging with other galaxies.

▽ Astronomers searching for the dark matter in our Galaxy in 1993 saw a distant star "flash" brighter over a period of a few weeks in just the way they expected if the light had been focused by a dark object acting as a gravitational lens. The time intervals between the images are 33 days, 12 days, 6 days, and 19 days. In 1994 about 50 events of this kind were detected.

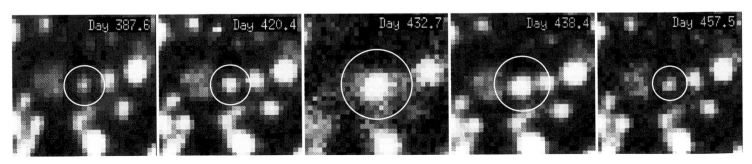

The COBE satellite has shown that very large-scale structures were already present when the universe first became transparent. In other words, the matter was already beginning to pull itself together into lumps and clumps. These concentrations became smaller in size and denser as gravity pulled the matter together. The concentrations seen by COBE may be the formation stage of the great superclusters of galaxies.

Merging galaxies

By using the Hubble Space Telescope and even natural gravitational lenses, galaxies in formation can be glimpsed. It seems that not all galaxies formed at the same time. The massive elliptical galaxies were formed shortly after the Big Bang, when the universe became transparent. Galaxy collisions were frequent, because the universe was smaller and more crowded. Galaxies grew in size by collisions and by merging with each other. Stars were formed at a furious rate, with frequent supernova explosions adding to the fireworks show. Today the elliptical galaxies have hardly any gas left in them, and most of their stars are very old.

Spiral galaxies seem to have formed later than elliptical galaxies. Around some of the original galaxies, disks of matter accumulated, and these later became the spiral arms.

In the Andromeda Galaxy, the Hubble Space Telescope has identified two separate nuclei, as if that galaxy formed as a merger of two smaller galaxies. We also know that our Milky Way is in the process of capturing the Large Magellanic Cloud. So it appears quite likely that galaxies grow and evolve through mergers and collisions. This is rare in the universe today, but observations of remote galaxies are showing us that there is a lot of activity in the young parts of the universe.

The future universe

Will the expansion of the universe eventually stop, or will it go on forever? At present, astronomers cannot say for certain which way the universe will go, although most of them believe the expansion will stop. The basic question is whether there is enough gravity in the universe to slow the expansion down and perhaps eventually reverse it. We are still far from firm conclusions, and the future of the universe depends very much on how much mass it contains altogether.

The evidence for dark matter

Is what we see through telescopes all there is? Probably not. Matter that astronomers believe exists but which they cannot detect is known as dark matter. Astronomers first speculated on the existence of hidden matter in the universe 60 years ago. For the next 40 years most astronomers were content to note this as a mere curiosity, but nevertheless the evidence that our survey of the universe is incomplete has gradually increased.

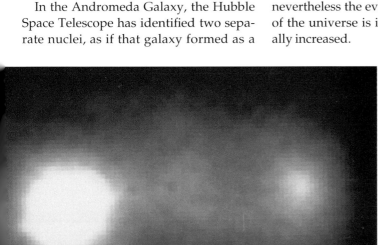

◁ This Seyfert galaxy has two bright nuclei in its central regions. This is very unusual, but the structure could have been formed if two galaxies merged without their separate nuclei fully blending together in the center.

In the case of our Galaxy, the high speeds of some stars and the orbit of the Large Magellanic Cloud both suggest that there is something more than stars lurking in space. Gravity is tugging so hard at the stars that the matter we see can account for only 10 percent of the mass of the Galaxy.

The sums do not add up in other galaxies either. Just as the orbital periods of the planets depend on the mass of the sun, so the speeds of stars in a galactic disk depend on the mass concentrated at the nucleus. From 1974 onward, astronomers found increasing evidence for unseen matter in spirals.

Moving up to the next level of size in the universe, we come to clusters of galaxies. Here the problems hit the crunch point. Once again, by balancing the forces acting on the galaxies, we can weigh the whole cluster. The masses are from 10 to 20 times higher than those we would jot down simply from looking at the light from stars in the galaxies. The conclusion is that clusters of galaxies contain a great deal of matter that does not emit any light.

The mass of the universe

What is the grand total mass of the universe? We can scarcely try to explain how the universe works unless we know how much of it there is. The mass density of the universe tells us right away whether it will expand forever or collapse back on itself. To weigh the universe at large, we use the speeds, or redshifts, of galaxies. The streaming motion of galaxies as part of the universal expansion gives us the mass of the whole universe.

Visible galaxies get nowhere near weighing down the expanding universe: they contribute about 5 percent of the ballast needed to anchor the universe from runaway expansion.

The hunt is now on to search for the dark matter that is weighing the universe down. There are many forms that this could take. Most astronomers believe it must be in the form of subatomic particles of some kind, and these would of course be very difficult to detect. They would not be a part of our everyday experience, and other astronomers wonder if the theory is correct. Perhaps, after all, this universe is unique and will go on expanding infinitely in the future.

STAR CHARTS

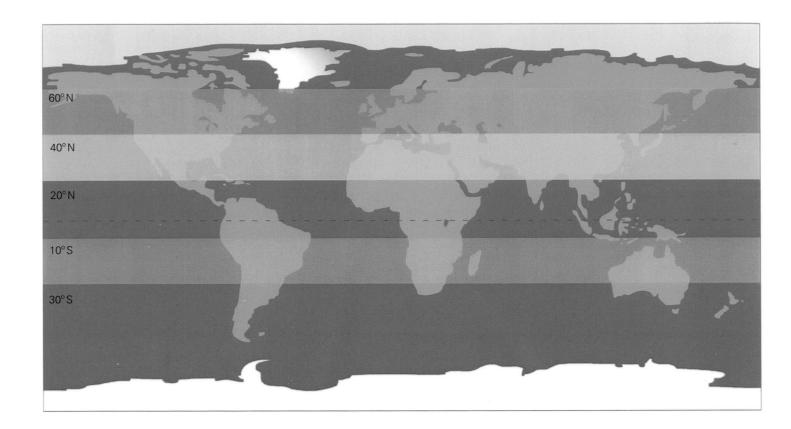

△ The three colored bands on this map of the world show the latitude bands in which each of the three sets of star charts on pages 151–153 should be used. The pages are color-coded along the edge to match the colors of the bands.

To start finding your way around the constellations, the first step is to identify a few really bright star patterns. Then you can use the constellations you know as stepping-stones to fainter, more difficult ones. If you are in a place with many street lights, or there is a Full Moon, you will only be able to pick out the very brightest stars. It will be easiest if you can start with a good clear, dark sky, but it is still worth trying for the brighter constellations even if you are in a city.

The stars you can see change from hour to hour through the night, and from night to night if you observe at the same time. The constellations you are able to see also depend on the latitude of the place where you are observing (see pages 26–29). For all these reasons, you need to make sure you have chosen the right star charts for your location and the time when you are looking.

On the next three pages you will find charts for three different latitude bands, which are marked on the map of the world above. First, with the help of the map, decide which latitude band you are in, then turn to the appropriate page of star charts.

Each pair of semicircles shows for various times and dates the stars you can see when you stand facing north or south. Above each of the pairs you will find the months of the year and the times of night when the constellations appear as shown. The four pairs of charts correspond roughly to the evening sky in each of the four seasons—winter, spring, summer, and autumn.

Choose the charts for the month and time combination most appropriate to when you are observing. You will need to add one hour to the times shown by the charts if daylight saving time (summer time) is in effect. If you are out observing in the early hours of the morning, you can still use the charts. They apply two hours later at night for each month earlier you observe, following the pattern of the times listed.

The larger the dot on the star chart, the brighter the star. The names of some of the brightest are marked. The Latin constellation names are in capital letters. The black lines are to help you look for patterns that tend to stand out. The lighter-colored band is the Milky Way.

LATITUDES 40°– 60° NORTH

Winter sky

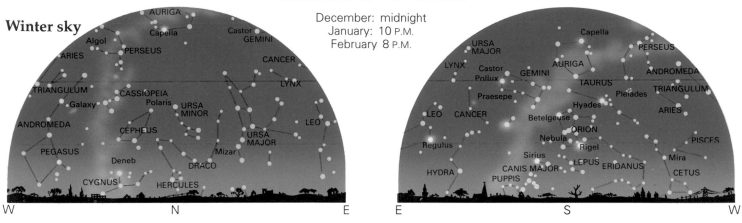

December: midnight
January: 10 P.M.
February 8 P.M.

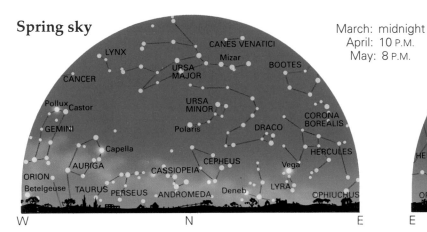

Spring sky

March: midnight
April: 10 P.M.
May: 8 P.M.

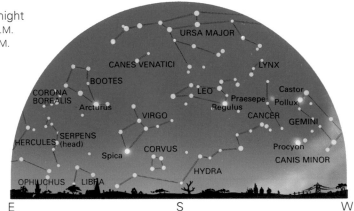

Summer sky

June: midnight
July: 10 P.M.
August: 8 P.M.

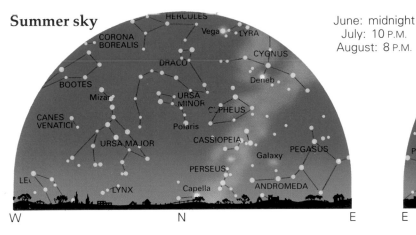

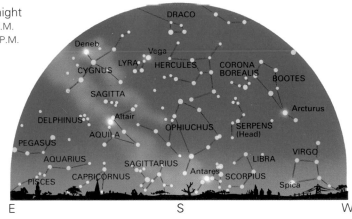

Autumn sky

September: midnight
October: 10 P.M.
November: 8 P.M.
December: 6 P.M.

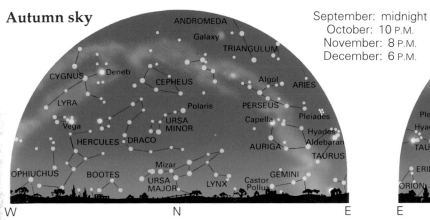

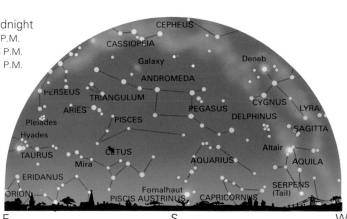

LATITUDES 20° – 40° NORTH

Winter sky

December: midnight
January: 10 P.M.
February 8 P.M.

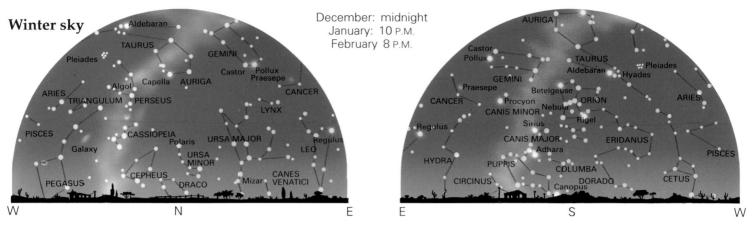

Spring sky

March: midnight
April: 10 P.M.
May: 8 P.M.

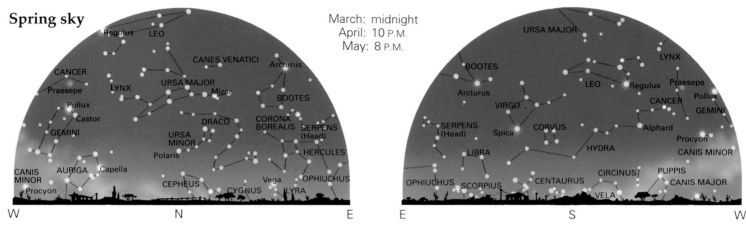

Summer sky

June: midnight
July: 10 P.M.
August: 8 P.M.

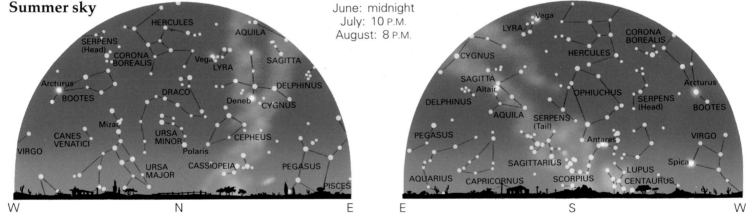

Autumn sky

September: midnight
October: 10 P.M.
November: 8 P.M.
December: 6 P.M.

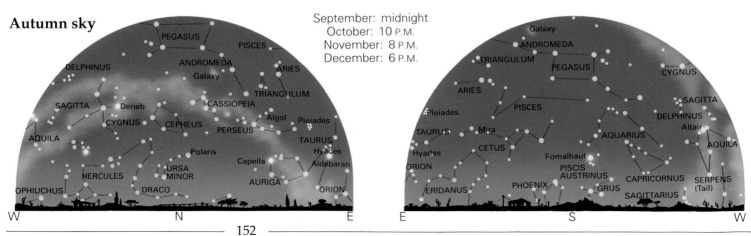

LATITUDES 10°– 30° SOUTH

Summer sky

December: midnight
January: 10 P.M.
February 8 P.M.

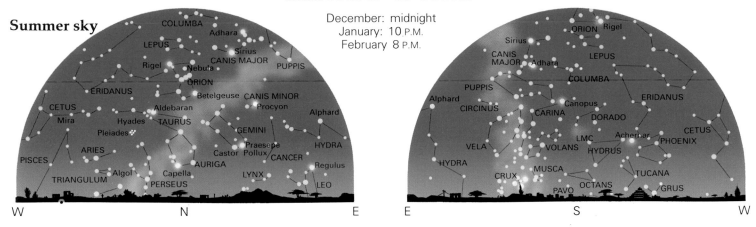

Autumn sky

March: midnight
April: 10 P.M.
May: 8 P.M.
June: 6 P.M.

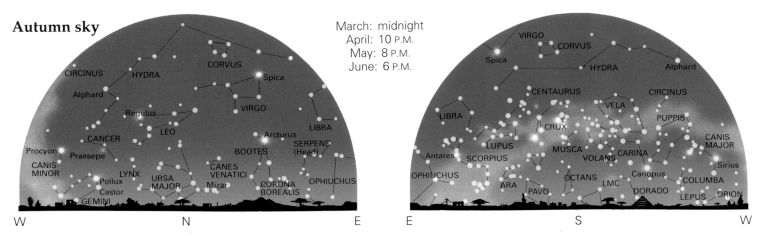

Winter sky

June: midnight
July: 10 P.M.
August: 8 P.M.

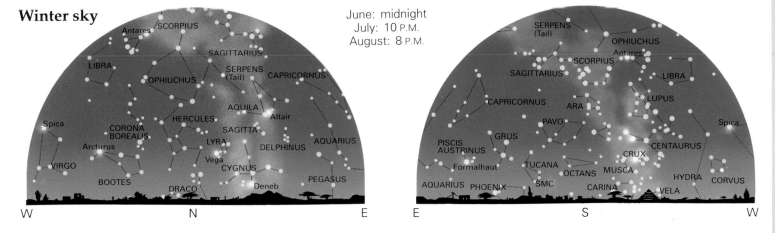

Spring sky

September: midnight
October: 10 P.M.
November: 8 P.M.

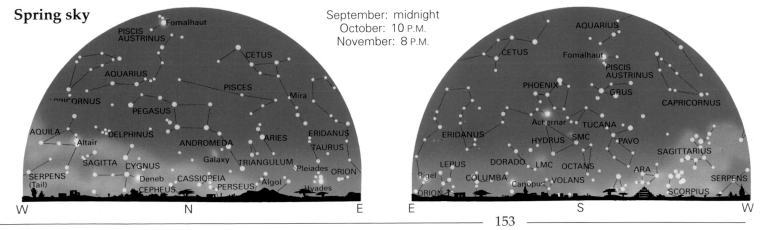

GLOSSARY

accretion disk A disk of material that collects around a spinning star.

active galaxy A galaxy emitting from its central region huge amounts of energy which is not coming from stars.

asteroid A piece of rock and/or ice orbiting the Sun like a tiny planet.

astrology An ancient tradition linking people and events with the positions in the sky of the Sun, Moon, and planets.

astronomical unit The average distance between the Earth and the Sun, 92,955,730 miles.

atmosphere The outer layers of gas surrounding a planet, moon, or star.

atom A tiny particle of matter consisting of a nucleus surrounded by a cloud of electrons.

aurora Colored glows in the Earth's atmosphere seen from time to time in the night sky over polar regions.

Big Bang A theory of the origin of the universe that says that it exploded from something incredibly tiny and hot and has been expanding ever since.

binary star A pair of stars in orbit around each other.

black hole A region in space where so much mass is concentrated that its gravitational pull prevents even light from getting out.

brightness The strength of radiation emitted by or received from a shining astronomical object.

brown dwarf A ball of material resembling a small dim star, but not massive enough to become a real star.

CCD Abbreviation for charge-coupled device, which records images electronically.

celestial sphere The sky imagined as if it were projected onto a huge distant sphere surrounding the earth, ignoring the real distances of the stars.

Cepheid variable star A type of star that varies in brightness over a few days in a regular way as it pulsates.

chromosphere The layer of gas around the Sun immediately above the visible disk.

circumpolar star A star that is always above the horizon as seen from a particular place and that circles one of the celestial poles as the Earth spins on its axis.

comet An icy object orbiting in the solar system that evaporates gas and dust and grows a tail when it is near enough to the Sun's light and heat.

constellation One of the named areas into which the whole of the sky is divided, or the pattern of stars in it.

Copernican theory The idea that the Sun and not the Earth is the center of the solar system, as established by Nicholas Copernicus in 1543.

corona The thin, very hot outermost layers of the Sun visible only during a total eclipse of the Sun.

cosmic background radiation Radiation left over from the earliest moments of the universe, which now fills all space.

cosmology The study of the universe as a whole.

crater A bowl-shaped feature on the surface of a planet or moon.

dark matter Unseen matter in the universe, known to exist because of its gravitational pull on other things.

day The time it takes the Earth to spin once, relative to the Sun.

declination The equivalent of latitude for specifying positions on the sky.

Doppler effect The change in the pitch of a sound or the color of light when the source is moving toward or away from the observer.

ecliptic The Sun's yearly path around the sky as seen from the Earth, or the Earth's orbit around the Sun.

electromagnetic radiation A form of energy that travels through empty space at the speed of light.

electromagnetic spectrum The full range of electromagnetic radiation spread by wavelength. From shortest to longest wavelength, the main categories are gamma rays, X-rays, ultraviolet radiation, light, infrared radiation, microwaves, and radio waves.

electron A tiny subatomic particle with a negative electrical charge.

ellipse An oval shape like an evenly squashed circle.

equinoxes The two times in the year (around March 21 and September 23) when the Sun is overhead at the equator at midday.

escape velocity The minimum velocity a spacecraft must reach to escape from the gravitational pull of a planet.

galaxy A large family of stars in space held together by gravity.

gamma rays The most powerful form of electromagnetic radiation.

gravity The attractive force that acts between all masses.

greenhouse effect The heating of a planet's surface and atmosphere when heat radiation from the Sun gets trapped by the gases in the atmosphere.

helium The second-lightest chemical element. It is rare on Earth but accounts for about a quarter of the material in the universe as a whole.

Hubble's Law The law that states that galaxies are all moving apart and their speeds increase in proportion to their distances.

hydrogen The lightest and simplest of all chemical elements. It accounts for about three-quarters of all the material in the universe as a whole.

infrared radiation Electromagnetic radiation we sense as heat with wavelengths in a range longer than visible red light.

interstellar medium The gas and dust between the stars.

Kepler's laws Three rules describing the motion of planets in their orbits around the Sun worked out by Johannes Kepler.

Kuiper Belt A region of the outer solar system, beyond the orbit of Neptune, populated by icy objects with the potential to become comets.

latitude Distance north or south of the equator measured as an angle.

lava Hot, molten rock that pours from an erupting volcano and turns back into solid rock after it has cooled down.

light The type of electromagnetic radiation that can be seen with the naked eye.

light pollution Artificial light that makes it difficult to see the night sky.

light-year The distance that light travels through empty space in one year—6 million million miles.

Local Group The small cluster of galaxies, with just over 30 known members, to which our own Milky Way galaxy belongs.

longitude Distance east or west of the zero longitude line, measured as an angle.

luminosity The amount of energy radiated per second by a glowing object.

Magellanic Clouds Two small galaxies neighboring the Milky Way and visible to the naked eye in the southern sky.

magnitude The brightness of a star or other astronomical object. The smaller the magnitude, the brighter the object.

maria (pronounced "mar-ree-uh") The Latin word for *seas*, used to describe the large dark areas on the Moon, although in reality they are solid rock. (The singular is *mare*, pronounced "mar-ray.")

meteor The bright trail in the sky of a small rock from space burning up in the Earth's atmosphere.

meteorite A piece of rock and/or metal from space that lands on the surface of the Earth (or another planet).

meteoroid A small, rocky object in space with the potential to become a meteorite if it hits the Earth.

meteor shower Meteors radiating from one point in the sky as Earth goes through a cloud of dust in space.

Milky Way The galaxy of stars to which our own Sun belongs, visible on dark nights as a hazy band of light encircling the sky.

minor planet An alternative name for an asteroid.

molecule A tiny particle of a chemical substance, made up of two or more atoms.

moon Earth's only natural satellite. Also, a natural satellite of any other planet.

nebula A cloud of gas and/or dust between the stars or around stars. Galaxies were called nebulae before they were known to be made of stars.

neutrino A subatomic particle with no electric charge and almost no mass, which travels practically at the speed of light.

neutron A subatomic particle with no electric charge, found in the nuclei of atoms.

neutron star A collapsed star in which all the atomic particles have squashed together to make a dense pack of neutrons.

nova A binary star that suddenly increases in brightness when material from one star cascades onto its partner.

nucleus The central core of a galaxy or of an atom.

objective The main light-collecting lens in a refracting telescope.

Oort Cloud A supposed cloud of potential comets totally surrounding the solar system at a distance of about one light-year. It has never been seen directly.

orbit The path through space of a body under the influence of the gravity of another object.

parallax The change in relative positions of objects at different distances when they are viewed from different places.

parsec A unit of distance measurement used only by professional astronomers, equivalent to 3.2616 light-years.

phase The amount of the surface of the Moon (or planet, etc.) lit by the Sun.

photon A particle or "packet" of electromagnetic radiation energy.

photosphere The visible surface of the Sun (or of any other star).

planet A ball of rock or gas orbiting the Sun, or another star, which is itself too small to become a star.

planetary nebula A star surrounded by a shell of glowing gas that it has blown off itself.

positron A small subatomic particle similar to an electron but with a positive electric charge.

prominence A flamelike jet of hot gas rising from the Sun's surface.

proton A subatomic particle with a positive electric charge, found in the nuclei of atoms. A single proton forms the nucleus of a hydrogen atom.

protostar A star in a very early stage of formation.

pulsar A neutron star beaming a rapid stream of radio pulses.

quasar An extremely luminous, remote galaxy that looks almost starlike.

radar Bouncing radio signals off objects to find their distances and shapes.

radio galaxy A galaxy emitting exceptional amounts of energy in the form of radio waves.

radio astronomy The study of the universe through the radio waves emitted by planets, stars, the gas between stars, and galaxies.

radio waves Electromagnetic radiation in the range that carries the least energy and has the longest wavelengths.

red giant An old star that has increased greatly in size and has a relatively cool surface that glows red.

redshift The increase in wavelength of light (or other electromagnetic radiation) when its source is traveling away from the observer.

right ascension The equivalent of longitude for describing positions on the sky.

satellite A natural moon of a planet or a spacecraft in orbit around a planet.

second of arc A unit of measurement for very small angles. There are 3,600 in one degree.

sidereal time Time measured according to the rising and setting of the stars, rather than the Sun. Astronomers find it useful for planning observations.

solar system The Sun and its entire family of planets and other objects (comets, asteroids, moons, dust, etc.).

solstices The times in the year when the Sun is overhead at midday at either the most northerly point (about June 21) or the most southerly point (about December 21).

spectroscopy Breaking down light, or any electromagnetic radiation, to study different colors or wavelengths individually.

spectrum The rainbow of colors and its extension to electromagnetic radiation at longer and shorter wavelengths, invisible to the eye.

star A large, glowing ball of gas generating its own energy by nuclear reactions taking place at its center.

sunspot An area on the Sun's surface that looks dark because it is cooler than its surroundings.

supernova The catastrophic explosion of a star, which can cause it to shine for a few weeks as brightly as a whole galaxy.

supernova remnant A shell of gas blown off by a supernova.

telescope In astronomy, any instrument for collecting radiation from any part of the electromagnetic spectrum.

ultraviolet radiation Electromagnetic radiation with wavelengths in a range shorter than visible violet light. This form of radiation causes sunburn.

universe Everything that exists.

variable star A star that changes in brightness over time, either regularly or in an unpredictable way.

white dwarf An old collapsed star that has run out of nuclear fuel at its center and is in the process of dying.

X-rays A powerful form of electromagnetic radiation with wavelengths in the range between ultraviolet radiation and the even more powerful gamma rays.

year The time it takes the Earth to orbit the Sun once.

zodiac The band of constellations around the sky through which the Sun gradually moves over the course of a year.

INDEX

Where several page references are given for a particular headword, the more important ones are printed in bold (e.g., **86–87**). Page numbers in italics (e.g., *94*) refer to illustrations and captions.

ACKNOWLEDGMENTS

Design: Raynor Design

Abbreviations: t = top; b = bottom; l = left; r = right;
c = center; back = background

Photographs
The publishers would like to thank the following for
permission to reproduce the following photographs:
Adventure Photo:
  A. P. Kennan Ward: 99b
Anglo-Australian Observatory: 4bl; 5bl; 38t; 92t; 94-95b;
  121t; 121b; 123tl
  Royal Observatory Edinburgh: 25l; 25r; 47t; 94l; 94-95t
  Telescope Board: 26-27; 103c; 137tl; 137cl; 138tl
AURA Incorporated, Kitt Peak National Observatory: 121t
Bodleian Library: 36tr
Britstock-IFA:
  Eric Bach: 36tl
  T. S. Chanz: 44tl
  Bernd Ducke: 8t
  Grahammer, Ducke: 126t
  H. Schmidbauer: 136t
*California Institute of Technology, Las Campanas
  Observatory*:
  J. T. Trauger: 72b
Collection Ville de Bayeux: 82-83t
M. Covington: 9br; 45; 103t; 109r
J. Dragesco: 59t; 96-97t; 97 inset
European Space Agency: 19r
Mary Evans Picture Library: 12bl; 40tr; 74bl
J. Fletcher: 115c
Galaxy Picture Library: 10t; 18
Hale Observatories: 121c
H. R. Hatfield: 32t; 34-35b
P. Janssen: 97b
Jet Propulsion Laboratory: 22t
W. M. Keck Observatory:
  Keith Matthews, James Larkin: 142tl; 142br;
  KPNO Composite: 142bl
Serge Koutchmy: 5bc; 26; 98r
Lick Observatory, Santa Cruz: 58; 116t; 116c; 118b
MACHO Project:
  Wil Sutherland: 148b
McDonald Observatory, Texas: 41
Mittons: 8cr; 118t;
  J. Mitton: 14tl; 19l; 23c; 23b; 36b; 55t; 65t; 74tl

S. Mitton: 6-7; 96bl
M. Moberley: 69b
NASA: 8bl; 8-9c; 17tl; 30-31; 51; 59b; 60t; 60b; 66b; 67t;
  67b; 68; 69t; 70-71t; 73; 74-75t; 74-75c; 75c; 75b; 75r;
  76c; 77t; 77 inset; 77b; 78b; 78-79c; 78-79b; 79tr; 86t;
  87; 96br; 98-99; 110b; 132-133b
Eric Bach: 82t; 96t
Matt Bobrowsky (CTA Incorporated): 123tr
*California Institute of Technology and Carnegie
  Institution of Washington*: 24-25
Carnegie Institution of Washington, Alan Dressler: 47b;
  142tr
COBE Science Working Group: 92-93; 126bl
European Space Agency: 72t; 80b; 98l; 123br; 143br;
Goddard Space Flight Center: 144
COBE Science Working Group: 92-93; 126bl
IRAS: 92b
Jet Propulsion Limited: 30tl; 54-55t; 54-55b; 56; 57t;
  57c; 57b; 57br; 65b; 66tl; 66c; 86t
Tod R. Laver, Sandra M. Faber: 137cr
Rice University, C. R. O'Dell: 103b
Space Telescope Science Institute:
  *Francesco Paresce, European Space Agency, Paola
  Sartoretti, University of Padua*: 74bl
  *R. Jedrzejewski, Francesco Paresce, European Space
  Agency*: 116b
  JHU, H. Ford: 141tc
  J. Mackerty: 149
  H. A. Weaver, T. E. Smith: 85b
US Geological Survey: 64t
National Optical Astronomy Observatories: 8-9b; 47c; 130t;
  132t; 138bl; 138br; 141tl; 141bl
  Steward Observatory: 12-13
  *University of Arizona, Gary Berstein, Tony Tyson
  (A. T. & T. Bell Laboratories)*: 143
National Radio Astronomy Observatory: 17bl; 123bl; 138tr
AUI: 136-137t; 138tr; 141br
  VLA: 137b
*National Research Council of Canada, Dominion
  Astrophysical Observatory*: 148t
Natural History Museum, London: 88t; 88br
Jack B. Newton: 8-9t; 102tl; 108; 121b; 130bl; 130br; 131b;
  133br
Palomar Observatory/Carnegie Institution: 6-7; 109l
Max Planck Institute Fur Aeronomie: 85t
The Planetarium, Armargh, Northern Ireland: 82-83b

T. Platt: 71
Royal Astronomical Society, London: 114t; 114c; 114b
Royal Greenwich Observatory: 14-15t; 15tr
Royal Observatory, Edinburgh: 102tr; 110t
  AAT: 128; 129l; 129r
John Sandford: 32b; 88-89t
Scott Sandford: 89br
UK Schmidt Telescope Unit: 133t
Science Photo Library/Hale Observatories: 46b
Scrovegni Chapel, Padua: 83t
Space Frontiers Ltd: 34-35t; 72c; 75l
  H. J. P. Arnold: 30tr
Space Telescope Science Institute:
  Dana Berry: 112t; 113b; 116-117; 126br; 136b; 141tr;
  143tl
  Antonella Fruscione, Richard Griffiths, John MacKenty: 16t
*UCLA,Astronomy Department, Ben Zuckerman and Harland
  Epps, Jon Gradie-Joan Hayashi, Planetary Geosciences
  Division, University of Hawaii, Robert Howell,
  Department of Physics and Astronomy, University of
  Wyoming*: 105l; 105r
University of Hawaii at Manoa, Institute for Astronomy:
  81 inset
University of Hawaii, Alan Stockton: 137t
University of Michigan: 1; 17r
University of Oxford, Department of Astrophysics: 134-135

Illustrations and diagrams
Julian Baum: 15c; 15b; 16b; 20; 22-23; 32-33; 40t; 44b; 47b;
  81; 84b; 86b; 108; 115t; 120; 126-127; 131; 138c; 139;
  144; 144-145
Paul Doherty: 10-11; 46t; 61t; 61b; 102b; 104; 111
David Hardy: 2; 4tl; 5br; 26-27; 28b; 30b; 37; 38-39; 40b;
  42; 43b; 48-49; 50t; 50-51; 52t; 52-53; 54tc; 58-59; 62-
  63; 64-65; 70t; 70-71; 73r; 74tr; 76t; 76b; 76-77; 78t; 78-
  79; 80t; 80-81; 84t; 90-91; 100t; 100b; 101; 106; 107;
  119r; 119b; 124-125; 146-147; 151-153
David Mallott: 66–67
Oxford Illustrators: 14; 18; 34t; 44tr; 97
Raynor Design: 23t; 28tl; 28tr; 29; 31; 38; 40b inset; 43t;
  47t; 88; 112b; 113t; 114b; 115b; 140; 150t

Cover
Front, back, and spine illustrations: Julian Baum
Background photograph: Lick Observatory, Santa Cruz